BASIC WATER
TREATMENT

BASIC WATER TREATMENT
for application world-wide

George Smethurst
BSc, FICE, MIWEM, FGS

SECOND EDITION

THOMAS TELFORD
LONDON

Published by Thomas Telford Ltd, Thomas Telford House, 1 Heron Quay, London E14 9XF

First published 1979
Second edition 1988

British Library Cataloguing in Publication Data
  Smethurst, George
    Basic water treatment for application world-wide.—2nd ed.
    1. Water supply. Treatment
    I. Title
    628.1'62

ISBN 0 7277 1331 0

Typeset, printed and bound in Great Britain by Redwood Burn Limited, Trowbridge, Wiltshire

Preface to the second edition

The reception given to the first edition of this book has been gratifying, particularly in developing countries where much water-works construction is proceeding because of the pressing need. In spite of this, there are more people at the present time who do not have access to water of reasonable quality than was the case ten years ago. This is because the population of the world increases by about a million people every five days and it has not proved possible to build new installations fast enough, or to operate the existing ones well enough, to keep pace with the ever-increasing demand for water.

The need for new works will continue far into the foreseeable future and a high proportion of this growth will occur in relatively hot countries. Treatment plants designed to normal UK standards work very well in warmer climates, provided that a few fairly obvious pitfalls are avoided. This book is designed to restate the basic principles of water treatment and to indicate the modifications sometimes required to meet unusual overseas conditions.

Preface to the first edition

This book has evolved from a paper on the influence of various climates and conditions on the performance of different types of settling basin. The trouble was to contain the subject. Local conditions are affected not only by rainfall and temperature, but by living and educational standards, economics, customs—all sorts of things. Settling basins are merely a part of the whole water treatment process. Their performance depends on what has happened to the water before it reaches the basins and what remains to be done after it has left. By the time full justice had been done yet another book on water treatment had emerged, and still the need remained to draw the line somewhere. The treatment of water is so vast a subject, and has attracted so much attention, that there is no point in rehashing material that has already been presented, except to the extent that is relevant to some specific aspect of the matter under discussion. More new waterworks are being built now than ever before, most of them in countries where money is needed for other things, and expertise is in short supply. There is a pressing necessity throughout the world for simple works which local people can operate and afford. It is this theme that this book pursues. A general knowledge of water engineering among its readers is assumed; given that, this book should help them to solve most treatment problems they are likely to encounter.

Contents

1: Introduction

The search for new sources of water is never-ending, particularly in newly emerging countries. Populations are increasing, people are getting piped supplies, and nobody wants to carry water from the village pump any more. In many towns in Asia and elsewhere the demand for potable water doubles every 10–15 years, not only because of the rising domestic consumption but to meet the needs of industry, and all these new supplies need treatment to a greater or lesser extent. Crops may need vastly greater quantities of irrigation water and thus compete for the water available, but treatment is rarely required for water used for agricultural purposes.

Because of the steepness of the demand curves, and the tendency of the steepness to increase, engineers everywhere have sought to curb consumption per head, but without success. Living standards are rising and people are demanding more sophisticated sanitary arrangements, better housing and bigger gardens. Forward predictions of demand invariably turn out to be too low; there seems to be nothing one can do about it.

All waterworks leak to some extent and these unaccounted-for losses may be quite serious. The most rigidly controlled Western cities are lucky if leakage can be reduced to 10% of the daily output, and where control is ineffective 25–50% loss is not unknown. It is impossible to reduce these losses except by inspection up to and within the houses and this may be difficult. In some countries access is forbidden to non-related males by religious custom. Possibly female inspectors may be the answer, but loss of water by leakage will always be a problem.

In tropical countries water consumption tends to be high. An average demand of about 300 l/head per day is not uncommon even in developing countries, and if there are big seasonal temperature variations, summer peak demands can be even higher.

People cannot exist without water, so there has always been a demand, and all the nearest and more obvious sources have already been exploited. Any future demand must inevitably be met from sites which are more remote or increasingly less attractive, and for this reason alone development costs must rise continuously in future. Coming at a time of inflation, present day costs are leaping upwards. Wages and conditions vary from country to country, but to give some idea of the size of the problem a typical undertaking, sited on a river and including a treatment plant, might cost about $500 per cubic metre per day of output capacity, which allowing for growth represents $330 per present head of population at 1988 prices. To pay for this, water charges are levied which should cover running costs and annual repayments. Waterworks are operated on commercial lines and if bankers lend money they want a reasonable rate of return, which may be about 8%, but may vary upwards or downwards depending on the source.

The communities which are building their first waterworks or first extensions normally have other equally important uses for such capital and individual skills as they possess. There is therefore a pressing need for works that can be built as cheaply as possible and operated by fairly unskilled labour. Compared with the difficulty of finding the initial capital, running costs seem of lesser importance and in some areas labour-intensive works might even be welcomed. Most plants in undeveloped countries tend to be too sophisticated and an all-too-common sight on any waterworks is some automation or electronics which has not worked since the day it was installed. This does not necessarily mean that the plant is not working, because most waterworks are large, good-natured things which if properly proportioned and reasonably well looked after will do a decent job. It simply means that money has been wasted.

The important thing is to concentrate on the essentials. By doing this it is possible to lower the price and produce good works which relatively unskilled people can handle.

It should not be assumed from the foregoing that there is no place for automation and sophistication in water engineering. In some countries they are highly desirable, if not absolutely essential. Even in many others they might be completely suitable for the bigger cities. But there are vast areas of the world where they should be avoided, certainly for the time being. The question that should be answered is whether the equipment can be effectively maintained and repaired. All too often the conclusion must be reached that only the more robust and simple items will remain continuously serviceable and works should be built with these limitations in view.

Most waters have to be purified before they can be used for potable purposes. Raw water is so infinitely variable in quality that there is no fixed starting point to the treatment process, and within much narrower limits there is no rigidly fixed finishing point either. Many countries have their own standards of acceptable purity for potable water, and these vary. The World Health Organization lays down two standards which are widely applied in developing countries (cf. Chapter 2).

There is virtually no water that is impossible to purify to potable standards, but some raw waters are so bad as to merit rejection because of the risk and expense involved. If a good quality source is not locally available, a second-class source may have to be upgraded by treatment to first-class standards, or better water may have to be brought in from some more distant source. It is generally a matter of economics, in considering which the prosperity and sophistication of the community served has to be taken into account.

Different countries have different economic structures. The amount of expertise, the cost of labour and the availability of materials are infinitely variable. There are climatic problems and even political ones. It follows that there is no one method of purifying water which is universally applicable, and this applies not only overall but to any of the various processes which combine to provide full treatment. This is particularly true in the case of the settling basins, of which there is a wide choice. Most of these have their merits and weaknesses and their own particular spheres of suitability.

Full treatment of water may comprise pre-treatment (with or without chemicals), mixing, coagulation, flocculation, settlement, filtration and sterilization. All are important but none more so than settlement, which removes up to 90 % of the suspended solids and can affect the performance of the whole works probably more than any other single process. Even so, not all waters require full treatment and not all treatment plants require settling basins. In any given case the amount of treatment required has first to be decided before consideration is given to the best way of providing it. There can be few engineering operations in which pure theory has more often to be modified by practical considerations, often dictated by local experience and particularly by the need to suit the capability of the people who are going to operate the plant. In many countries the typical plant operator is not very highly paid and high class performance cannot be assumed. Highly skilled plant chemists who are readily available in big Western cities may be non-existent in many parts of the world and sophisticated plants may be at a disadvantage.

The quality of the raw water is always difficult to foresee. No matter how many water samples are examined prior to plant design, the probability that the worst possible conditions will have been discovered is not high. Quite apart from seasonal variations in the raw water quality, the possibility of radical long term changes due to development in the catchment area or the building of impounding reservoirs should be borne in mind. A clear stream emerging from a jungle-covered catchment can turn into an extremely turbid one if part of the jungle is cleared for development at some later date.

River water can also change its chemical and biological character if it is impounded. As an example of this, the water stored in the Kalatuwawa reservoir which serves Colombo developed some extremely troublesome characteristics after impounding commenced. Not only did the water in the lower zone of the reservoir become deficient in dissolved oxygen and high in soluble iron and hydrogen sulphide, which happens in most deep reservoirs, but the top 10 m of the water developed a heavy crop of algae of the *Stephanodiscus* type. These are notorious filter-blockers and they caused recurrent trouble at the Kalatuwawa treatment plant until steps were taken to control them.

Any works built on the lower reaches of a river may, by increased abstraction, cause salt intrusions to travel further upstream at times of high tide, making the raw water more saline.

Practical problems of this kind are mostly to be found when treating water from surface sources. If future deterioration of the raw water is considered to be likely, either the proposed plant should be designed to cover against the worst foreseeable contingency or, at least, provision should be made to add future treatment capacity as and when required.

Water from underground sources is generally clear, but it may be excessively hard, or contain iron and manganese. The soluble salts of calcium and magnesium, which cause hardness, and also those of iron and manganese can be thrown into suspension by chemical reactions and, thereafter, removal can proceed by settlement and filtration in much the same way as silt can be removed from river water. In fact the process is normally a little easier to control because groundwaters tend to remain more uniform in quality than river waters and the precipitates formed are much heavier and settle more quickly than does the finer sediment typically found in river-derived supplies.

In tropical and subtropical countries climates may be wet, as in Nigeria and Brazil, or dry, as in North Africa and the Middle East. Rivers flowing through arid countries are prone to carry very high silt loads and this creates a need for bigger settling basins than would be found in the UK. Rivers in the wetter tropics tend to carry a lot of floating debris and weed, which make elaborate screens essential. All screens need constant cleaning, and in wet and humid parts of the world particular attention must be paid to the efficiency of the methods by which the floating material can be removed from the screen after being intercepted.

While most overseas waterworks will be built where climatic conditions are as warm as and probably a lot warmer than in the UK, a smaller number will be required where it is a good deal colder. Low temperatures do not suit most water treatment processes, and settling basins might have to be rather bigger (see Chapters 7 and 8) to compensate for this. In addition, where the average temperature during the coldest month is below 0°C it is advisable to enclose the filters and sometimes the settling basins to avoid freezing. In this connection it should be remembered that some countries, like Iran, which one thinks of as 'hot' have a short but bitterly cold winter.

In areas where seismic shocks are experienced, there is no additional treatment problem but all structures have to be designed to withstand earthquakes. Most countries have their own building regulations and codes of practice in this regard. There is a general belief that water-retaining structures which are circular in plan resist earthquakes better than those which are rectangular, and one often has a choice. Records of all earthquakes are kept at the Institute of Geological Sciences, Edinburgh.

In desert countries where water might be scarce, brackish, or even non-existent, it is often necessary to resort to desalting sea water. Ships at sea have long used the waste heat from their propulsion machinery to distil fresh water from salt. The design of land-based plants spread rapidly after the Second World War, a system known as flash distillation came into use, utilizing the waste heat from power stations, the water 'flashing off' under high vacuum to produce several times more product water than the weight of steam applied. All forms of desalting are expensive and troublesome and distillation is now being overtaken by a process known as reverse osmosis (RO) which depends on pumping salt water at very high pressure (up to 60 bar) through special membranes which have the capacity to permit water to pass through them but not salt. RO is in a very active stage of development and progress continues to be made in perfecting and lengthening the working 'life' of the membrane. RO is generally considered now to have overtaken distillation as a means of desalting brackish water and is certainly comparable even on sea-water installations.

When comparing distillation with RO, the different quality of the end product should be kept in mind. Water from an RO plant has high dissolved solids, while distilled water is virtually solids-free. However, the latter can be very corrosive and both need to be sterilized; in distillers the water boils at very low temperature due to the high vacuum so sterilization due to boiling does not occur. Both types of plant require highly skilled design and operation and their great cost prohibits use except in places like oilfields, where the wealth generated and the availability of imported technicians make them viable.

A system known as electrodialysis can be used with advantage on water which is not very brackish (with total dissolved solids below about 7000 mg/l). Electrodialysis depends upon the basic principle that negative ions are attracted to a positive electrode and vice versa. The salt in brackish water can be concentrated into a brackish stream and ejected, leaving relatively fresh water to go into service.

Desalting in any form should be regarded as a method of last resort. Except in the most extreme circumstances it is seldom a practical option.

2: Quality of water and treatment required

ORIGIN AND TYPE OF IMPURITIES

Absolutely pure water is rarely found in nature. The impurities occur in three progressively finer states—suspended, colloidal and dissolved. Different methods of treatment are required for their removal or reduction to acceptable limits.

Suspended matter

Quite apart from the probability that a river might be carrying floating debris, running water has the capacity to pick up and transport solid particles of higher density than water, the higher the velocity the bigger the particle picked up. Rivers in flood are therefore normally at their most turbid because of the increased velocity in the channel. Transportation of the coarser solids occurs at the velocities shown in Table 2.1. Big rivers in flood often run at velocities in excess of the maximum shown in Table 2.1 and are capable of carrying heavy loads of sediment.

Table 2.1. Transportation velocities of particles (adapted from Fox[1])

Material	Diameter of particle, mm	Velocity of water, m/s
Fine sand	0.4	0.15
Medium sand	1.1	0.22
Coarse sand	2.5	0.30
Gravel	2.5–25	0.75
Shingle	25–75	1.20

Colloids

The finer particles, colloids, may not be visible to the naked eye but in their finest forms impart colour to the water. Colloids remain in suspension even when the water is virtually at rest.

Dissolved solids

In its passage over or through the ground, water also picks up such substances as are soluble, notably calcium, magnesium, sodium, potassium, iron and manganese; in their soluble forms, these are normally combined with bicarbonates, sulphates, chlorides, nitrates and other salts. Gases also may be absorbed, particularly carbon dioxide, oxygen, nitrogen and ammonia. Few dissolved solids are particularly objectionable in low concentrations.

Organic pollution

Pollution from organic sources is regarded with much more concern. The products of decomposition of organic wastes such as nitrates and nitrites may be regarded as indicators of pollution. The presence of *Escherichia coli* (*E. coli*) provides positive proof of such.

Tastes and odours

Under certain conditions algal growth may be present (cf. Chapter 3), and if so may impart objectionable tastes and odours to the water. The removal of algae is essential and often a little difficult.

Certain elements such as phenol also cause taste even when present in incredibly low concentrations of as little as one part in 10–20 million parts of water.

Hardness

The bicarbonates, sulphates and chlorides of calcium and magnesium commonly found in water cause hardness. Hardness forms insoluble precipitates with soap which leads to waste. It also causes boiler scale.

Iron and manganese

Iron and manganese, if present in any concentration above the very slightest, impart tastes, and stain articles which are being washed.

Sulphates, chlorides and fluorides

The sulphates of magnesium and sodium if present in excess act as laxatives. Chlorides in concentrations above 600 mg/l tend to give the water a salty taste. Fluorides in concentrations above 1.5 mg/l are undesirable and in concentrations above 3 mg/l may cause mottling of teeth.

Detergents and insecticides

Detergents and insecticides (cf. Chapter 12) are apt to find their way into raw water and are objectionable if present in excess.

Corrosiveness

Corrosiveness is a characteristic of waters with high CO_2 in conjunction with a low pH value and low alkalinity. As a result, water softened by the base-exchange process is apt to attack metals, as may water from a distillation plant. Among natural waters, soft, peat-stained moorland waters are generally corrosive.

POTABLE WATER STANDARDS

Any of the conditions mentioned above may be present in any given raw water and many of them certainly will be present. Although there are exceptions, surface waters tend to be high in suspended solids and low in solids in solution, whereas the opposite is more typical of groundwaters.

There are no hard and fast rules as to the acceptable quality for potable supplies but certain guidelines have been laid down. If these are not exceeded no action is necessary, because the cost of providing and operating treatment plant is appreciable and may represent a significant item of expenditure among low-income groups. The World Health Organization is actively concerned in spreading piped water supplies in developing nations and has set two international standards for the impurities still present after treatment: the Highest Desirable and the Maximum Permissible (Table 2.2). Inevitably both relate to water of lower quality than that which would be demanded by richer and more sophisticated communities, but they are realistic in those areas for which they are intended. In Western countries the 'high class' standard of Table 2.2 is often attained or exceeded, but

Table 2.2. Standards for potable piped water supplies

Substance or property	Upper limit in high class water	WHO Highest Desirable concentration[2]	WHO Maximum Permissible concentration[2]
Total solids, mg/l	300	500	1500
Colour, units on platinum–cobalt scale	3	5	50
Turbidity, Jackson turbidity units	0.2	5	25
Taste	0	*	*
Odour	0	†	†
Iron, mg/l	0.1	0.3	1
Arsenic, mg/l	0.05	0.05	0.05
Manganese, mg/l	0.02	0.1	0.5
Cyanide, mg/l	0.1	0.2	0.2
Copper, mg/l	0.2	1	1.5
Zinc, mg/l	1	5	15
Lead, mg/l	0.05	0.05	0.05
Calcium, mg/l	‡	75	200
Magnesium, mg/l	‡	50	150
Sulphate, mg/l	‡	200	400
Chloride, mg/l	100	200	600
pH	7.0–8.3	7.0–8.5	6.5–9.2
Magnesium and sodium sulphate, mg/l	200	500	1000
Phenol, mg/l	0	0.001	0.002
Carbon chloroform extract (CCE) (organic pollutants), mg/l	0.05	0.2	0.5§
Alkyl benzyl sulphonate (ABS) (surfactants), mg/l	—	0.5	1
Hardness (CaCO₃), mg/l	80	200‖	300‖

* Unobjectionable (threshold odour number (TON) <2).
† Unobjectionable (threshold odour number (TON) <2).
‡ Not stated.
§ Concentrations above 0·2 mg/l require further investigation to determine causative agents.
‖ No recommendation made, but the values shown would be reasonable.

as water is infinitely variable in quality, certain constituents may exceed the values shown even in otherwise first-class supplies.

The standards in Table 2.2 relate to the final quality of water supplied. Except for certain borehole supplies, a water as good as the 'high class' water would almost certainly have been filtered.

TURBIDITY

Turbidity is caused by the presence of suspended solids in the water. The latter is a measure of the total weight of dry solids present whereas turbidity is an optical effect, stated in turbidity units, which also reflects the fineness, colour and shape of the dispersed particles. For this reason there is no constant linear relationship between the two, but the two values are often numerically of the same order and tend to be used indiscriminately.

There are several methods of measuring turbidity, and the results are frequently expressed in different units. In the past this led to considerable confusion but most modern units are numerically the same and the confusion is subsiding. Care should be taken in interpreting any results stated in parts per million on the old silica scale because these can vary quite appreciably according to the type of silica used to make the suspension.

Methods of measurement

The turbidity rod consists of a rod marked off in centimetres, at the lower end of which two bright platinum wires are extended at right angles. One of these is of 1 mm dia. and the other of 1.5 mm. When immersed in water and viewed from above there is a depth at which one wire disappears while the other remains visible. The depth of immersion can then be related to the turbidity by direct reading from Table 2.3. This is an extraordinarily useful piece of equipment to carry about even though the results given are only approximate.

There are a number of types of turbidimeter, one of which is the Jackson turbidimeter. In this a standard candle is viewed through a column of water under test, the length of which is increased until the flame disappears. The length of the column defines the degree of turbidity, which is stated in Jackson turbidity units. These pieces of apparatus are now calibrated by using suspensions of formazin polymer which are stated in formazin units. These units have been

Table 2.3. Measurement of turbidity by visibility of immersed wire (adapted from Suckling[3])

Depth of immersion, cm	Turbidity, parts per million on silica scale	Notes
2	1000	
4	360	Filters
6	190	rapidly
8	130	clog
10	100	
15	65	Filters operate with difficulty
30	30	Special operation
45	18	called for
80	10	Maximum desirable limit for water entering filters

adopted in the USA by the American Public Health Association as standard and they are now almost universally used.

Turbidity units

The nephelometric turbidity unit (NTU), the Jackson turbidity unit (JTU), the formazin turbidity unit (FTU) and the APHA turbidity unit are all numerically the same and are interchangeable for all practical waterworks purposes. It is also possible to make silica scale standards giving values in milligrams per litre which are numerically the same, but it should be assumed that many old references to this standard may not be identical. Owing to differences in the type of silica used and different laboratory techniques it may not always be possible to correlate old SiO_2 standards with new standards or even with each other. The old unit stated in parts per million on the silica scale represented the degree of turbidity caused by 1 mg of SiO_2 in 1 litre of clear water. It would appear to have had values varying between 1 JTU and 0.2 JTU. A type of silica—Fulbent 370—gave a silica scale/Jackson ratio of 4.8, but a ratio of about 2.5 was common in many UK laboratories.

In view of this the interpretation of Table 2.3 may be questioned but enquiries indicate that in this case the silica scale units used were somewhat similar to JTUs. As an aid to waterworks management, however, the value of the table cannot be questioned, whatever the units might be. It is valuable to a works superintendent to know instantly by putting his turbidity rod into the outlet channel from the sedimentation basins whether operating conditions are likely to be good, mediocre or potentially impossible at any particular moment and he will rapidly relate the immersion depths to whatever units he happens to be working in, and to his previous operating experience on the works for which he is responsible.

LIMITS OF TURBIDITY OF SETTLED WATER ENTERING FILTERS

Filtration is a final polishing process and there are limits of incoming turbidity beyond which filter performance deteriorates. To obtain really first-class water a filter has to be fed with water which has already reached a fair degree of clarity. In most plants 70–90 % of the clarification is done by the settling basins, leaving only 10–30 % of the work to the filters. Where the source is heavily turbid the removal of the turbidity assumes particular importance.

Not all types of filter are equal in performance. Within their various degrees of effectiveness, all filters can adjust their performance as the incoming water deteriorates, provided that in the earlier stages of such deterioration they are washed more frequently. If the incoming water continues to deteriorate the filter can still operate reasonably well provided that the water is passed through it more slowly. For every type of filter, however, there is a degree of turbidity beyond which it cannot effectively operate.

The standard rapid gravity filter (and also the pressure filter) works well when the water reaching it from the settling basins has a turbidity of less than 10 JTUs. For the best results, most waterworks operators would strive to get the turbidity of the settled water down to 2–5 JTUs where possible. However, this type of filter can produce a satisfactory filtrate if the turbidity of the water from the basins is as much as 20 JTUs, although with anything approaching this amount of silt to remove the intervals between washing become progressively shorter. Under emergency conditions, the filters, working at reduced speeds and with frequent washes, can produce reasonable filtrates with incoming turbidities up to 50 JTUs, but that should be regarded as the outer limit of their capacity. This figure was exceeded in Bagdad in 1952 and in Tehran in 1977 and in each case complete stoppage of the works occurred, and a restart was only possible several hours later when all the filters had been washed. The freak river conditions which caused both crises were of very short duration and by the time the filters had been washed the rivers were returning to more manageable conditions.

Mixed media filters can do approximately 50 % better than this. Although they are normally operated to give better results with well clarified water, they can work with turbidities up to 25 JTUs without too much difficulty and for 25–75 JTUs can produce acceptable results. For this reason they can sometimes be used without settling basins; this practice is possible on certain lake waters. They are also very effective where algae are a nuisance, because of the large storage capacity of the anthracite layer. However, they are more expensive than rapid sand filters and their adoption should be preceded by a careful economy study.

Slow sand filters are not now very common because they take up a lot of space and are labour-intensive, neither of which considerations might preclude their adoption in a developing country. Many people think that they still have a definite place in water treatment because of the high quality of the filtrate. They are ineffective in removing natural colour and excessive suspended solids but can work with incoming turbidities up to 30 JTUs. Unlike the two previous types of filter mentioned they are operated without chemical coagulants, a characteristic which makes it unlikely that they would be preceded by a settling basin because the typical modern clarifier depends very much on chemically assisted floc formation. Because of this they are usually found working on reservoir-derived waters, which normally have turbidities well below the critical limits as a result of the lengthy natural settlement period provided. Where they have been installed for many years, as in London, the high quality of their filtrate makes it unlikely that their use will be discontinued. Only a few new plants have been built in recent years, however.

In most cases filters cannot operate without assistance from other processes. The function of treatment provided prior to filtration (which generally means the settling basins) is to take the raw water as it occurs in nature and reduce the turbidity to a level that the filters can easily accept.

PERMISSIBLE QUALITY OF RAW WATER

There is almost no limit to the degree of turbidity which might be encountered in rivers in certain parts of the world, but in England suspended solids in excess of 1000 mg/l are usually confined to the River Severn and even there occur only for short periods. By comparison, in 1952, the River Adhaim in Iraq was calculated on one occasion to be carrying 70 000 mg/l suspended solids. In the same year the Kansas River at Kansas City was carrying 11 800 mg/l suspended solids, and for eight months the suspended solids did not fall below 1000 mg/l. In Tehran in 1977, the River Karaj was observed to be carrying 57 789 mg/l suspended solids on one occasion.

Table 2.4 shows typical figures observed during floods on other rivers. Such values might be expected to occur fairly frequently and do not represent the worst ever recorded.

It will be appreciated that no basin can continue to work at its full design rate and deliver water with a maximum turbidity never exceeding 25 JTUs and normally of 2–5 JTUs irrespective of the condition of the raw water. It is sometimes possible to slow down the rate of working, and thus effectively decrease the overflow velocity, to give the basin a better chance. The Tigris is a good river in this respect because maximum turbidity never coincides with peak summer output and treatment plants can be operated at reduced speed through the worst parts of the winter and spring floods. No river need be rejected because of a high silt load. It is simply a matter of installing big enough basins, or pre-treatment.

The presence of harmful bacteria in raw water is far more serious than silt. Although any water can be purified, the risk to public health and the cost and care entailed put practical limits on the sewage pollution considered acceptable. The degree of pollution of

Table 2.4. Suspended solids in foreign rivers[4]

	Dry silt by weight, mg/l
River Irrawady	10 000
River Nile	5 000
River Hooghly	3 600
River Jumna	4 000

Table 2.5. Raw water quality (based on table from American Society of Civil Engineers et al.[5])

	Excellent source	Good source	Poor source	Rejectable source
Average BOD (5 days), mg/l	0.75–1.5	1.5–2.5	2.5–4	> 4
Maximum BOD in any one sample, mg/l	1–3	3–4	4–6	> 6
Average coliform, most probable number (MPN) per 100 ml	50–100	100–5000	5000–20 000	> 20 000
Maximum coliform, MPN per 100 ml	Less than 5% of samples > 100	Less than 20% of samples > 5000	Less than 5% of samples > 20 000	—
pH	6–8.5	5–6, 8.5–9	3.8–5, 9–10.3	< 3.8, > 10.3
Chlorides, mg/l	< 50	50–250	250–600	> 600
Fluorides, mg/l	< 1.5	1.5–3	> 3	—

water is judged from certain items in the chemical and bacteriological analyses. Table 2.5 shows commonly quoted standards; these are rather conservative and can be exceeded slightly. A modern treatment plant with pre- and post-chlorination can, in fact, purify water with a coliform count up to 1 million in 100 ml, but this calls for expert control and unremitting vigilance.

SUMMARY OF PERMISSIBLE TURBIDITIES OR SUSPENDED SOLIDS

For a treatment plant to be fully effective the following values should not be exceeded:

raw water in river—suspended solids concentration can be anything but should preferably be < 1000 mg/l; if it habitually exceeds 1000 mg/l a pre-settlement tank is indicated

at outlet from pre-settlement tanks—suspended solids < 1000 mg/l at outlet from main settling basins—turbidity 2–5 JTUs; in dire emergency 25 JTUs

at outlet from filters—turbidity < 1 JTU.

TREATMENT REQUIRED

Earlier in this chapter it was convenient to describe the required condition of the water in its progress through the plant in reverse order: first the final filtrate, secondly the settled water and lastly the state in which it might occur in the river. In actual practice a plant has to work in the opposite direction (Fig. 2.1).

A treatment plant consists of many processes—screening, coagulation, flocculation, sedimentation, filtration and sterilization—each of which is intended to perform one main function although it may incidentally partially assist with some other. The impurities are removed in order of size, the bigger ones being eliminated first. Not every water contains all the impurities and therefore not every water requires all the treatment processes.

The impurities are mainly removed as follows:

floating objects by screening;

algae (if present) by straining;

excessive iron, manganese and hardness in solution by precipitation in basins after the addition of chemicals;

normal suspended solids by settling;

the remaining fines and some bacteria by filtration;

excessive bacterial pollution by pre-chlorination;

final bacteria surviving filtration by chlorination.

All these processes overlap to some extent and many need auxiliary processes to be fully effective. Table 2.6 indicates the treatment applicable in each case.

Settling basins and filters form the basis of all treatment plants and in view of their widespread use the question must arise as to whether there are any circumstances under which they can be omitted. They should be provided in all cases except the following.

(a) A naturally clear groundwater which contains no iron, manganese or hardness in solution ('clear' certainly means less than 5 JTUs, and probably much less) needs neither basins nor filters;

Table 2.6. Recommended treatment* (based on table from American Society of Civil Engineers et al.[5])

	Coarse screens	Fine screens	Microstrainers	Pre-chlorination	Raw water storage	Preliminary settlement	Aeration	Flocculation	Coagulants and settling	Rapid filtration	Mixed media filtration	Slow sand filters	Post-chlorination	Superchlorination and dechlorination	Lime (-soda) softening	Activated carbon	Special aids	Desalting
Floating debris	E	E	O															
Algae			O	O					E	E	O							
Turbidity:																		
0–5 JTUs									O	O	O							
5–30 JTUs	O	O						O	E	E	O	O						
30–100 JTUs	O	O						O	E	E							O	
100–750 JTUs	O	O						O	E	E							O	
750–1000 JTUs	O	O			O	O		O	E	E							O	
> 1000 JTUs	O	O			O	E		O	E	E							O	
Colour:																		
< 30 Hazen								O	O	O								
> 30 Hazen								E	E	E								
Tastes and odours							O									E		
Calcium carbonate > 200 mg/l															E			
Iron and manganese:																		
< 0.3 mg/l				O			O			O								
0.3–1 mg/l							O		E	E							O	
> 1 mg/l				E			E		E	E								
Chlorides:																		
0–250 mg/l																		
250–600 mg/l																		O
> 600 mg/l																		E
Coliform bacteria, MPN per 100 ml:																		
0–20													E					
20–100									E	E			E					
100–5000				O					E	E			E					
> 5000				E	O	O			E	E			E					

* E essential, O optional.

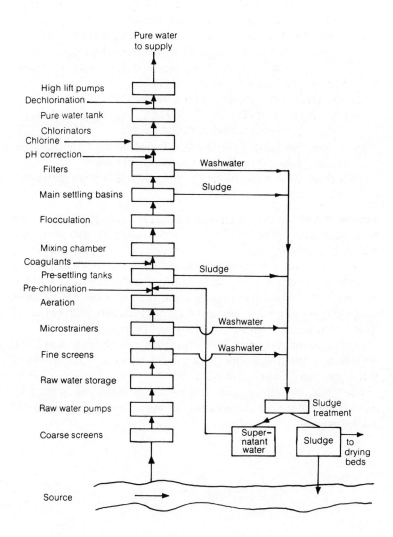

Fig. 2.1. Flow diagram showing possible treatment stages

(b) a reservoir- or lake-derived water of low colour with a turbidity less than 30 JTUs, the final treatment of which is on slow sand filters, does not need basins; nor does

(c) a reservoir or lake water of turbidity never exceeding 25 JTUs, the final treatment of which is on rapid sand filters (even so the omission of basins is far from universal because of the desirability of clarifying down to < 5 JTUs);

(d) any water with turbidity never exceeding 50 JTUs and habitually having turbidity of 5–10 JTUs, for which multi-layer filters are proposed, does not need basins;

(e) a clear, but hard, water being softened by the base-exchange process does not need basins or filters; nor does

(f) a clear water with colour less than 30 mg/l on the platinum–cobalt scale.

Settling basins should be regarded as very highly probable on all river abstraction schemes, and all lime softening and iron and manganese removal plants.

As regards filters, it is generally accepted that they can be omitted if the turbidity of the water being supplied does not exceed 5 JTUs, but this is not regarded as first-class practice and most communities would require a greater degree of clarification (to < 2 JTUs). If filters are used, even greater clarity would be expected.

In certain major cities (e.g., Rangoon, Perth and Adelaide), settlement without subsequent filtration is currently practised, but this is unusual in cities of this size.

The higher turbidities quoted above would be unusual in the UK, but are apt to happen occasionally on works treating highly turbid foreign waters.

3: Pre-treatment

It is not unknown for water to be pumped directly from the river to the basins but generally there are intermediate processes which can collectively be called pre-treatment. They include screening, raw water storage, pre-chlorination, aeration, algal control, straining, preliminary settling, coagulation, mixing and flocculation. Any of these processes might be found at a particular plant but it is improbable that all would be needed. Each performs a particular function and unless the problem they are designed to eliminate is present in the raw water they can be omitted.

INTAKE SCREENS

Coarse screens are provided at river intakes to prevent floating material of fairly large size entering the works. The steel bars forming the screen are normally quite substantial (about 25 mm dia.) and are spaced about 100 mm apart. They are often placed on a slight inclination from the vertical to facilitate raking. Sometimes the bars are mounted in frames which are duplicated so that one frame can be lifted for cleaning or repair without admitting unscreened water to the plant. The velocity of water through the screen openings should not exceed 0.5 m/s. Screens serve little purpose in rivers like the Tigris, which carries silt but not leaves or other floating debris. At Bagdad the ordinary strainers fitted to the raw water pump suctions provided completely adequate protection for the pump impellers. Rivers as free of floating material as the Tigris are few and screens should normally be regarded as essential.

RAW WATER STORAGE

In many instances and for various reasons, raw river water is stored for appreciable periods. Most pre-plant storage is provided to even out variations in flow. The storage may take the form of an impounding reservoir on a flashy stream, or a smaller open storage reservoir

Table 3.1. Quality of water before and after storage

	River Thames at Teddington[1]		River Thames at Oxford[2]		River Great Ouse at Diddington[3]	
	Raw river	Stored water	Raw river	Stored water	Raw river	Stored water
Colour, Hazen	830†	450†	19	9	30	5
Turbidity units (type not specified)	35	5.3	14	3.2	10	1.5
Chloride as Cl, mg/l	16.7	16.3	30	29	54	54
Oxidized N, mg/l	2.6	2.4	5.0	0.26	—	—
Ammoniacal N, mg/l	0.06	0.02	—	—	0.3	0.06
Albuminoid N, mg/l	0.15	0.10	0.19	0.2	0.5	0.4
Nitrite N, mg/l	—	—	—	—	0.1	0.01
Nitrate N, mg/l	—	—	—	—	2.0	0.3
Oxygen absorbed, mg/l	2.12	1.53	1.8	1.3	3.5	2.0
BOD, mg/l	—	—	—	—	4.5	2.5
Total hardness, mg/l	—	—	300	259	430	280
Iron, mg/l	—	—	0	0	0.1	—
Phosphate, mg/l	—	—	0.8	0.33	3.0	*
Total dissolved solids, mg/l	—	—	415	360	—	—
Alkalinity, mg/l	—	—	207	172	—	—
Presumptive coli, MPN per 100 ml	—	—	60 000	200	6500	20
Presumptive coli, % of samples						
of 100 cm^3	99.9	57.2	—	—	—	—
of 10 cm^3	97.7	32.5	—	—	—	—
of 1 cm^3	83.1	13.4	—	—	—	—
of 0.1 cm^3	48.3	3.3	—	—	—	—
E. coli, MPN per 100 ml	—	—	20 000	100	1700	10
Colony counts per 1 ml:						
3 days at 20°C	4465	208	—	—	50 000	580
2 days at 37°C	280	44	—	—	15 000	140

* Trace.　　† MWB Standard.

Table 3.2. Change in water quality due to storage

| | Reduction due to storage, % ||
	River Severn at Hampton Loade[4]	River Derwent at Draycott[5]
Colour	28	67
Turbidity	70	51
Chloride as Cl	0	0
Oxidized N	41	—
Albuminoid N	—	57
Nitrite N	—	45
Nitrate N	—	20
Total hardness	—	0
Iron	—	85
Manganese	—	67
Total dissolved solids	—	*
Presumptive coli	95	99
E. coli	94	> 99
Colony counts per 1 ml:		
3 days at 20°C	88	—
2 days at 37°C	89	—

* Negligible.

to tide over periods when the rate of abstraction at a river intake has to be diminished or suspended.

In all such cases an improvement in the raw water quality is noticeable purely as a result of retention of the water in the reservoir. It therefore follows that improvement in quality of a very low grade water can be achieved by storage alone, even if such storage is not required for any other purpose. This improvement results from natural settling of the suspended solids, which is accompanied by a marked reduction in pathogenic organisms. It is recommended that storage provided purely to improve quality should be equivalent to 7–15 days of the average water demand. This is sufficient to reduce pathogenic bacteria and river algae, while at the same time not long enough to encourage other organisms to develop.

However, storage is apt to cause a silt problem. It is not easy to install scrapers in large open reservoirs and, in the absence of some sort of silt trap, these relatively small reservoirs through which raw water passes fairly quickly tend to silt up.

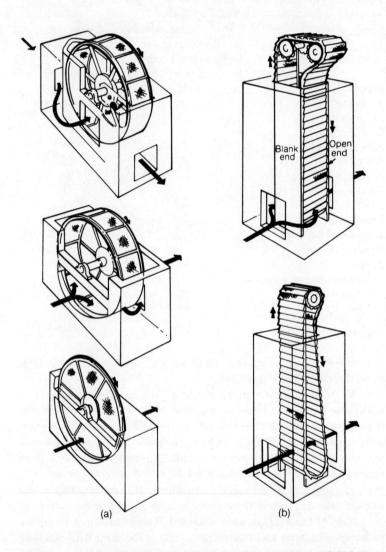

(a)   (b)

Fig. 3.1. Mechanical screens (by courtesy of Messrs Blakeborough Ltd): (a) mechanical types (drum; cup; disc); (b) travelling or band types (twinflow; uniflow)

The cost of building such reservoirs is also fairly high and they should be omitted unless the quality of the raw water falls below that of a 'poor source' as listed in Table 2.5. The high cost is caused by the fact that it is usually necessary to build on flat and valuable riverside land, on which a reservoir can be formed only within a low earthen embankment which completely encircles it. The maximum water depth of various known examples is about 15 m. Undue shallowness encourages weed growth and should be avoided. If it is geologically possible for the materials which are needed to form the embankment to be 'borrowed' from within the reservoir basin, some additional depth and reduction of cost can be effected. It is impossible to give an accurate figure but an approximate 1988 price of $4 per cubic metre of water stored might be of the right order of magnitude.

The beneficial effect of storage on water of low quality is shown in Tables 3.1 and 3.2. These examples have been selected at random from various works on fairly polluted rivers in England where storage has been provided. Perhaps the reduction in harmful bacteria is the most striking feature, giving point to the old waterworks saying: 'A stored water is a safe water.'

FINE SCREENS

Fine screens are normally fitted immediately after the coarse screens except where raw water storage is provided, in which case it might be more advantageous to place them at the outlet from the storage reservoir. The orifices in a fine screen mesh are of the order of 6 mm dia., and tend to clog very quickly if there is much floating debris other than silt. It is possible to fit the fine mesh in frames and remove these for cleaning by hand but this is only practicable when the raw water is carrying little debris, or in locations where labour is cheap and plentiful. It is now common practice to install one of the many proprietary forms of mechanical screen which are constructed on the endless band or drum principle and cleaned continuously by water jets which wash the strainings away along channels. The wash-water has normally been treated and must therefore be included in the plant losses, but this rarely amounts to more than 1 % of the total works throughput.

The screens are rated in terms of the amount of water they can pass at lowest working water level. They can be purchased in several different sizes and are housed in chambers through which the

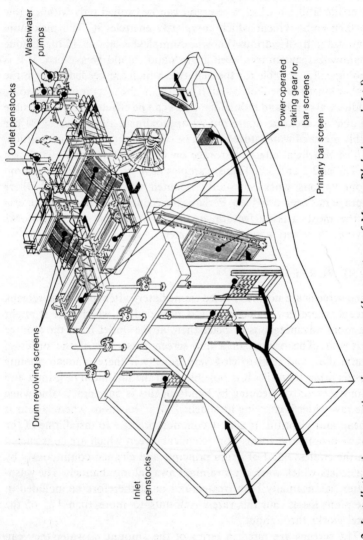

Washwater pumps

Outlet penstocks

Power-operated raking gear for bar screens

Primary bar screen

Drum revolving screens

Inlet penstocks

Fig. 3.2. Installation incorporating drum screens (by courtesy of Messrs Blakeborough Ltd)

Table 3.3. Fine screens

Type	Location	Dimensions	Rated capacity, gal/h
Drum type	Cardiff	30 ft dia. × 9 ft 3 in	1625 000
Cup type	Cornwall	12 ft dia. × 2 ft	135 410
Disc type	Lancashire	14 ft dia.	260 000
Uniflow band	France	4 ft wide × 19 ft 2 in	352 204
Twinflow band	South Africa	4 ft 3 in wide × 24 ft	2083 333

incoming water must pass. Typical examples are listed in Table 3.3. Typical installation details are shown in Figs 3.1 and 3.2.

PRE-CHLORINATION

Pre-chlorination refers to the practice of injecting chlorine into the raw water soon after it is abstracted from the river. On waters of reasonable quality this step is generally omitted. It will normally be most effective where the raw water is not particularly turbid but has a high bacteria count. Fairly high doses of chlorine (2–5 mg/l) are commonly used; during the lengthy period spent by the water in the settling basins, this oxidizes and precipitates such iron and manganese as might be present, kills algae and bacteria, reduces colour and slime formation and generally assists settlement.

If excessive silt is present in suspension the merit of pre-chlorination declines, as silt absorbs chlorine without compensatory benefit. It is not very effective, therefore, for heavily turbid water.

On clear groundwaters with a high ammonia content it can be used to advantage.

A drawback to using pre-chlorination is that raw water has a high chlorine demand and much greater quantities of chlorine are absorbed than in later chlorination to effect the same degree of sterilization. It should not be regarded as a substitute for post-chlorination, however. It is essentially an additional safeguard to be adopted only when extremely polluted (but clear) raw water has to be used.

The effectiveness of chlorine is reduced by low temperatures but this hardly affects pre-chlorination because of the long period of contact during the passage of the water through the plant.

AERATION

Aeration provides oxygen from the atmosphere to effect beneficial changes in the raw water. At the same time it may liberate undesirable gases such as carbon dioxide and hydrogen sulphide. The absorption or release of gas in water is a slow process unless the water is agitated or a great area exposed to the atmosphere. This is commonly done by splashing the water over trays to break up the stream into countless droplets or by reversing the effect and blowing air bubbles through the water.

Gases are absorbed or liberated from water until equilibrium is reached between the natural content of each gas in the atmosphere and its content in the water. Thus if the water is lacking in dissolved oxygen it will pick it up from the air, and its taste will improve; if it has an excess of carbon dioxide or hydrogen sulphide it will tend to lose them, with even greater benefit to the control of taste and corrosiveness. As hydrogen sulphide is non-existent in the atmosphere removal tends to be complete, whereas for more complicated reasons the removal of carbon dioxide may only be partial. Other substances of low volatility may react less favourably to aeration.

Aeration is therefore a cheap and valuable means of controlling tastes, odours and corrosion but not in all cases may complete control be achieved.

Aerators are commonly found to be necessary if any of the following conditions are present in the raw water:

(a) hydrogen sulphide (tastes, odours etc.);
(b) carbon dioxide (corrosive tendencies);
(c) tastes due to algal growth (caused by volatile oils released);
(d) iron and manganese in solution;
(e) de-aeration.

The action in cases (c)–(e) is to increase the oxygen content, and in (a) and (b) to liberate excess gas. Both are natural processes which the aerator effectively stimulates.

Hydrogen sulphide is often found in water from the lower depths of an impounding reservoir. It imparts a characteristic 'bad egg' odour to the water which is readily removed by aeration, particularly when the pH is low. However, there is a tendency for sulphur to be precipitated, which creates a chlorine demand, and sulphates might be formed which tend to be reduced back to hydrogen sulphide in the stagnant conditions of dead ends in the distribution system.

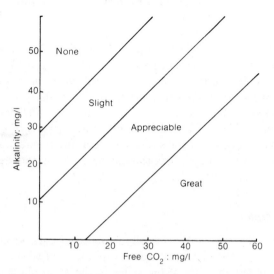

Fig. 3.3. Aggressiveness of water to concrete structures

Carbon dioxide is less easily reduced and never completely eliminated. Unless the water is naturally alkaline, the remaining CO_2 encourages corrosion and lime may have to be added even following aeration to combat this tendency. Fig. 3.3 indicates the corrosive effect on concrete to be expected at various CO_2/alkalinity ratios; in this regard it should be assumed that aeration will not reduce a high CO_2 content below a final concentration of 4.5–10 mg/l. The effect of CO_2 on ferrous metals is more severe and a different CO_2/alkalinity ratio is necessary (cf. Figs 13.1 and 13.2). The effect on concrete is worth recording separately; concrete used in the basins and filters can be seriously affected as the semi-treated water is passing through the plant. The increased alkalinity which may be required in the end product can be provided in the final stage of treatment, to the extent necessary.

Certain volatile substances liberated by algal growths or decomposing organic matter can be released from water by aeration, but this is only one step in dealing with algal problems, which are more fully discussed in the next section of this chapter. Aeration is not always effective in reducing tastes from this cause. In the case of phenols it might actually intensify the nuisance. (Phenols are not caused by algae but can cause taste problems in concentrations as low as one part in ten million.)

Iron in excess of 0.3 mg/l and manganese in excess of 0.1 mg/l are objectionable in water. They impart taste and colour and stain laundry. If present they are normally in solution but are precipitated by oxygen added either by aeration or chlorination. Aeration is generally used and it is effective in the majority of cases. The effectiveness depends on the way in which the iron and manganese compounds are combined. Cox[6] lists ten different methods of reducing the iron and manganese concentrations according to circumstances. Six of the most common incorporate aeration in the treatment process.

Types of aerator

Aeration can be effected by specially designed nozzles which direct thin jets of water against metal plates to produce a fine spray which exposes countless droplets of the water to the atmosphere. Nozzle type aerators are very efficient and are commonly used in the CO_2 and iron removal processes. The nozzles are commonly of 2.5–4 cm dia. and discharge about 5–10 l/s. To aerate 10 000 m³/day about 20 nozzles arranged within an area of 25 m² are needed. Greater or lesser amounts can be treated proportionately. Care must be taken to shelter the installation against wind, or the fine spray may be blown away from the collecting trays. The area where the sprays are working is often sheltered by louvres set in a surrounding wall.

Cascade type aerators depend on the turbulence created in a thin stream of water flowing swiftly down an incline and impinging against fixed obstacles. The surface area of liquid exposed is rather limited and there is a loss of efficiency.

Tray type aerators consist of about five perforated trays, increasing in size from top to bottom. The water falls from tray to tray through a total depth of 2–3 m and splashes on, through and off the trays 5–6 times. The total area of the trays in relation to the flow is generally about 0.1 m² per m³/h. This type of aerator is apt to freeze in cold weather and to encourage the growth of algae and other life in warmer climates. However, it is a simple and cheap method, used much more widely than any of the others.

Diffused air aerators consist of tanks in which air is bubbled upwards from diffuser pipes laid on the floor, the air holes in the diffuser pipes being sufficiently fine and numerous to promote a cloud of small bubbles. This type of aerator is efficient because bubbles tend to attain a constant terminal velocity whereas falling droplets tend to accelerate, and therefore for a given depth the air–water contact in a diffused air plant is longer. Also the amount of air used can be regulated according to need. Aeration tanks are commonly about 4 m deep and have a retention time of about 15 min. The air blowers should deliver about the same amount of air in any given time as the throughput of water, at a pressure of 5 psi. For a throughput of 10 000 m^3/day an air blower of about 5 hp is needed.

Relative effectiveness. Assuming diffused air to be the most effective method of aeration, a very approximate measure of relative effectiveness, measured in terms of CO_2 removal effected by a diffused air installation, would be

sprays	90%
trays	70%
cascades	50%.

Trays are most commonly used because of their cheapness, simplicity and reasonably high efficiency.

ALGAE AND ALGAL CONTROL

Algae are minute organisms which are usually classified as plants and proliferate in rivers and reservoirs. There are a multiplicity of types, many of which, if present to excess, can cause trouble to waterworks operators. Algae can normally be seen in affected water, but some (notably *Stephanodiscus*) can remain invisible except under a powerful microscope. Outbreaks tend to be severe and sporadic and sometimes coincide with a seasonal rise in temperature.

Some parts of the world are more prone to outbreaks than others and it is unlikely that any engineer would be entirely unaware of potentially serious algal trouble before becoming involved in the design of a works. The building of an impounding reservoir on an otherwise clear stream may encourage their growth, particularly in the upper layers of water. Otherwise fairly alkaline waters containing appreciable concentrations of nitrates and phosphates are particularly suspect. Clarity is also a factor because algae need sunlight for the photosynthesis of carbon dioxide. Problems caused by excessive silt and outbursts of algae rarely coincide, therefore. Likewise, heavy pollution inhibits the growth of algae by reducing the oxygen content of the water, but these conditions may disappear downstream as the stream recovers, and outbursts of algae may reappear. At Asunción in Paraguay the river water is completely clear of algae but the changed conditions within the water treatment plant encourage algal growth in the basins and filters. Some deep well waters from the chalk in southern England can be seriously affected when exposed to sunlight in open tanks, even for a short time.

The south Midlands of England are very prone to this nuisance, and serious outbreaks have been observed in Singapore, Kalatuwawa (Ceylon), Asunción (Paraguay) and Lake Ontario, to name but a few of the hundreds of affected areas. In other parts of the world algae are either absent or give little trouble. The River Tigris is almost completely free and is typical of other rivers in Iraq and Iran in this respect. This is probably due to silt and the absence of algal nutrients in the arid catchment areas.

Local knowledge is therefore essential because the algal problem may be anything between non-existent and completely crippling.

Control

Algae tend to float and are therefore not the easiest of materials to remove by means of settling basins. Where settlement is attempted the basins which have proved most successful are upward-flow tanks incorporating the sludge blanket principle, and it is desirable that the algae should be killed by pre-chlorination before reaching the basins. Chlorine doses of about 1 mg/l are normally fairly effective in killing the algae. The merit of flotation would appear to be obvious although there are as yet few examples.

Growth is inhibited by copper sulphate in concentrations of about 0.3 mg/l. It is usual to scatter this chemical on the surface of the water, particularly in shallow places, at a 'dressing' of 5 kg/ha, but this is not too easy to do in practice because the rivers or lakes in which the algae are prone to bloom are of such magnitude that dosing is difficult to control. Copper sulphate is also toxic to certain species of fish in the dosages required to inhibit algae and may not therefore be acceptable where fishing rights are important.

Where the organisms are visible and include *Daphnia*, *Cyclops* (Crustacea) or bloodworms, much heavier doses of copper (2 mg/l) or chlorine (3–50 mg/l) may be necessary, and these would require removal before the water could be used.

Strainers are widely used to remove algae, either in the form of rapid sand filters running without coagulants or as microstrainers. If the medium is coarse the former are sometimes known as roughing filters and can be worked at high speed to take the load off the main filters. Microstrainers are mainly of proprietary make and are capable of excellent performance provided that the water is relatively silt-free. The mesh of a microstrainer may have openings of about 20 μm and throughput velocities may be about 10 m/h through the submerged area. The filter elements are cleaned continuously by jets, as for a fine screen. The volume of washwater is 1–3% of the throughput. The loss of head is small (about 15 cm).

There are a number of different types of strainer, all working on the same principle. They suit some waters but not others; in favourable cases they will reduce the algal problem by 80–90%. The critical factor is silt. Microstrainers do not like silty water, but fortunately neither do algae. They perform admirably in areas where the water is clear and algal growth is heavy, such as the south Midlands of England and Lake Ontario.

PRE-SETTLEMENT BASINS

There is no doubt whatsoever that with heavily silted waters, big conventional horizontal-flow basins perform better than vertical-flow basins, and this undoubtedly accounts for their popularity in countries with rivers prone to excessive turbidity. It is common knowledge that upward-flow basins become increasingly difficult to operate once the dry silt by weight exceeds 1000 mg/l. Where this is likely to happen it is helpful to put in a small, non-chemically-assisted, horizontal-flow basin immediately upstream of the vertical-flow basins to keep the peaks of suspended solids well below 1000 mg/l. This is less essential with horizontal-flow basins, but it is desirable because it saves chemicals. The amount of coagulant used cannot be directly related to the amount of silt present in any given water, but the one does follow the other to some extent, and a small pre-settling basin will quickly pay for itself on turbid rivers.

Pre-settlement tanks

(a) permit the use of upward-flow basins on rivers formerly considered too turbid, if for other reasons their use is preferred;

(b) will rescue existing upward-flow tanks prone to get into difficulties (as at the Sarafiyeh works, Bagdad);

(c) may be built later if conditions on the gathering ground deteriorate (a common occurrence in developing countries where jungle-covered catchment is apt to be cleared for development without any notice or proper control);

(d) can be used with horizontal-flow basins to save coagulants;

(e) can be inserted to reduce the silt load on basins which are difficult to clean (this is one of their secondary functions at Tehran).

Early pre-settlement basins on the Missouri River were designed on the basis of 3 h nominal detention time and an overflow rate of 4–4.5 ft/h (about 30 m/day), largely in accordance with findings of Bull and Darby,[7] and these have been used on waters of up to 20 000 mg/l suspended solids with very great success. At Kansas City in 1952,[8] the suspended solids in the Kansas River had a peak of 11 800 mg/l, and for 8 months stayed over 1000 mg/l. The average suspended solids in the effluent from a 3 h uncoagulated preliminary basin was 320 mg/l and never rose above 600 mg/l.

However, Bull and Darby reached their conclusions 50 years ago, before treatment plants attained their present efficiency; it is now felt that 3 h detention is more than is needed, and many successful tanks with not more than 1 h of detention capacity have recently been constructed. Observations on 1 h capacity uncoagulated pre-settlement tanks in Bagdad and elsewhere confirm that effluents with considerably less than 1000 mg/l of suspended solids can safely be expected from such tanks, even though the incoming raw water might have suspended solids of 10 000 mg/l or more. The pre-settlement tanks at Tehran have an effective volume of less than 1 h capacity, but it took raw water with total suspended solids of 57 000 mg/l to overthrow them for a period of about 12 h, and this happened only once in 20 years' operation.

The 30–60 min capacity of pre-settlement tanks is positively lavish in comparison with grit chambers on sewage works, which often have peak flow retention periods of only 90 s with depths of 1.5 m and through flow velocities up to 30 čm/s. By comparison, in a square pre-settlement tank 3.5 m in depth, with an overflow rate Q/A of 80 m/day, the horizontal velocity would be about 0.5 cm/s and in practice most particles of diameter of about 0.1 mm and upwards would be deposited.

The falling velocities of siliceous sand in still water are given in Table 6.2, from which it can be seen that theoretically all particles above 0.03 mm in diameter would hit the floor of a tank 3.5 m deep within 1 h. However, conditions are never perfect in a tank and in practice somewhat larger particles would carry over, and would be settled out in the main basins.

Very exceptional cases

In certain parts of the world, notably Algeria and certain parts of Eastern China, quite phenomenal concentrations of silt have been observed. These have been known to reach levels of 200 000 mg/l, albeit for very short periods. Although it is true that any condition can be dealt with given a big enough basin, there is insufficient experience to give any guidance as to what size might be required to cope with water of this quality. Certainly, preliminary basins of 3 h capacity would be an absolute minimum, and one would try to discontinue operation if at all possible during the worst period. In such areas the use of horizontal-flow settling basins of very ample size to follow pre-settlement would also be advisable. One would be thinking in terms of not less than 6 h nominal retention time in the main basins, in which, of course, coagulants would have to be used. Fortunately, few engineers are likely to encounter silt concentrations of this severity!

Condition of water at inlet to main settling basins

Once the provision of pre-settlement tanks, where conditions require them, has been assumed, it follows that all the water reaching the main settling basins should have a suspended solids concentration of less than 1000 mg/l by dry weight. Under these conditions any properly designed and operated set of basins should be able to produce water fit to admit to rapid gravity filters. It therefore further follows that, from the outlet of the main settling basins onwards, any overseas waterworks can be designed in accordance with UK, European, or US standards.

4: Coagulants and coagulant aids

The vast majority of settling basins are dosed with chemicals capable of producing an adsorbent bulky precipitate. There are many substances which react suitably with water to produce such an effect and these are known collectively as coagulants. The precipitate so formed in the water is called the floc; the larger and heavier the floc, the quicker is the rate of settlement. It is frequently beneficial to use additional chemicals which, while not themselves true coagulants, intensify and improve floc formation. These are called coagulant aids.

Coagulants should be added immediately downstream of any pre-settling basin which may have been considered necessary. Their primary purpose is to assist in the removal of the more finely divided sediment and the colloids. Most of the larger and heavier particles settle unaided in the pre-settling basin, thus permitting the coagulants to work more efficiently on the finer particles. Thorough mixing is vitally important and flocculation may be equally desirable and therefore once the pre-settlement stage has been passed (if used at all) the coagulants must be introduced as soon as possible.

Most coagulants are salts of iron or aluminium and on mixing with water they act by a process of double decomposition involving the mutual interchange of groups, the end product being hydroxides in the form of gelatinous precipitates (the floc). In soft waters where there is insufficient alkalinity to react with the coagulants it has to be added either as lime or soda-ash. This serves to neutralize the sulphuric acid which forms, together with hydroxide, when sulphates hydrolyse. If left in the water the acid would recombine with the hydroxide and revert to sulphate. Hydroxide is the desired end product. It is insoluble, floc-forming and heavier than water, and it carries the positive electric charge necessary to neutralize the negative charges of the colloidal particles. Further reasons for the addition of some form of alkali are to establish the optimum pH value at which coagulation can take place, and to raise the final pH value after treatment to reduce corrosiveness.

Where floc formation is poor, or for reasons of overall economy, coagulant aids may be added. By producing a heavier, faster-settling floc, this allows smaller basins to be used, and smaller doses of the main coagulants may also be possible. The choice of the best coagulant for any particular water is determined by experiment.

Where the case for using a coagulant aid depends on cost-saving, the constantly changing nature of river water may increase the necessity for expert control. In developing countries the overall annual cost of providing an expatriate chemist, including air fares, family expenses, housing and other allowances, amounts in 1988 to at least $80 000, and this should be taken into account when steps involving increased sophistication are considered. It is a further example of the different economic backgrounds against which selection of process must be made.

Even one river may react differently to different coagulants at different seasons. Alternatively, it may be found that a dose of 1 mg/l of activated silica might equal in effectiveness a dose of 0.1 mg/l of polyelectrolyte at less cost. This sort of sophisticated juggling is practicable in countries like the UK but not necessarily in a developing country, where the knowledge of waterworks chemistry possessed by the operators might be limited.

It is important to differentiate between coagulation and the succeeding step of flocculation. Immediately coagulants are introduced into the water rapid mixing is essential. Floc starts to form and if, immediately after this, the water is stirred very gently, the fine particles adhere to each other and grow into settleable floc. This gentle stirring action occurs to some degree naturally in all basins but can be greatly accentuated by methods designed to encourage the rolling action required.

Of the four distinct steps in the addition of chemicals—preparation and dosing, mixing, coagulation, and flocculation—the first three invariably take place outside the basin proper. The last may or may not. Certain designs of basin are multi-chambered and therefore flocculation does in fact proceed more separately than might appear to be the case at first glance.

In coagulation, the pH of the water is important. Water is classified as neutral at a pH value of 7, which is the point where the weight of the hydrogen ions (H^+) equals the weight of the hydroxyl ions

(OH$^-$). Values of pH below 7 indicate acidity and those above 7 alkalinity. Floc formed in any given water tends to be heaviest at a specific pH value which has to be determined by use of the jar test (cf. Chapter 14).

Soft, coloured and acid waters with pH values of 5–6.5 are often difficult to clarify, needing treatment with alkali and coagulant at dosages which tend to be narrowly critical. Waters of pH 6.2–7 with a reasonable degree of alkalinity react well to aluminium sulphate. Alkaline waters with pH values of 7–7.8 may again be difficult and absorb high doses of alum.

Most coagulants are acidic and cause the pH value of water to fall on admixture. This may have to be corrected by dosing with alkali. Floc formation tends to be satisfactory below pH values of 6.8 and above 7.9. Some waters have a dead area between 6.9 and 7.8 where flocs are fine or non-existent. Each individual water has its peculiarities which may vary seasonally and require frequent adjustment. Any intelligent plant superintendent gets to know his own works, and is normally capable of carrying out the simple routine tests which are necessary.

Where low pH has been maintained in the basin to aid flocculation and lime is added finally to correct any corrosive tendencies of the treated water, it is vital that this should be done after filtration. All floc-forming reactions are reversible and if the pH is raised prior to filtration the floc will go back into solution and pass through the filter, carrying excess metallic salts into supply.

CHEMICAL DOSING PLANT

Coagulants can be added to the water either as a solution, which is much the commonest way, or in powder or slurry form. As treatment is a continuous process, dosing must also proceed in continuous and controlled fashion.

Solution feed

(The following remarks on solutions apply also to 'suspensions'.)

The two essential parts of a solution feed system comprise a tank in which a solution of the correct strength may be stored, and a dosing rate-of-flow controller. The tank should hold 24 h supply and be duplicated so that one tank may be in service while the other is being replenished. There should be some sort of continuous stirring mechanism to obviate the risk of settlement after initial preparation. Many coagulants (particularly alum and ferric chloride) are corrosive and the tanks should be lined with acid-resisting material, commonly rubber, glass or special cement.

The dosing mechanism should be capable of being controlled manually. There are two kinds of dosers: gravity-feed, and displacement pumps or tippers. The rate of flow can be altered in the former by altering the size of the outlet orifice in a constant-head tank, in the latter by altering the length of piston stroke of the specially made plunger pumps. The speed at which tippers operate can also be regulated. In big works of sophisticated design, dosing can be automatically controlled.

Dry feeders

A dry feeder incorporates a hopper which contains the powder and feeds a measuring device. This often takes the form of a revolving table from which a scraper of adjustable length deflects a greater or lesser amount of powder into the raw water. If the powder is not very soluble it may be mixed with water and fed as a slurry. In humid, tropical conditions, trouble by 'caking' is sometimes experienced on dry feeders and for this reason solution feeders are preferred.

Most mixers lend themselves to automation, with the rate of flow of coagulant dependent on the rate of flow of water through the works. On a small or unsophisticated works where the rate of flow tends to be constant, the simplicity of manual regulation is much to be preferred.

Dosing appliances are mostly of proprietary make and they vary widely in detail, their mode of operation being described in the various makers' literature.

STORAGE AND HANDLING OF CHEMICALS

In many Eastern countries chemicals have to be imported and arrive at the works in bags or drums. In larger cities it is increasingly likely that delivery will be make by bulk carriers designed to transport powders that may be unloaded pneumatically, or liquids which can be pumped.

On small-to-medium works in developing countries, however, large consignments of bags or drums carrying the coagulants have occasionally to be handled and stored. At typical dosage rates, even a moderately sized works of, say, 25 000 m³/day output would use over a tonne of chemicals daily; thought should therefore be given to initial off-loading, storage and daily transportation to the solution tanks, particularly when these are at high level. On all but the smallest works in countries where labour is cheap and plentiful, some sort of mechanical equipment is necessary. In its simplest form this may merely comprise hand trolleys and hoists, but in bigger installations highly sophisticated bulk handling machinery can be justified. Such machinery is rarely designed only for waterworks but is similar to that used in installations handling sugar, flour, lime or similar substances. The equipment is normally bought as a package either from waterworks plant manufacturers or from the makers of the individual items.

Strength of solutions

In mixing solutions of any chemical it should be noted that a 5% solution means that 5 parts of the chemical should be added to 95 parts of water (by weight) to get 100 parts of solution, and so on. An 8% solution would contain 8 kg of chemical to 92 kg of water. Percentages normally relate to the actual substance (e.g., alum, lime) being handled and not to any of the basic elements (e.g., calcium, aluminium) therein included.

ACTION OF COMMONLY USED COAGULANTS

The reactions of chemicals used as coagulants are quoted in many standard text books [1-4] and are reproduced here for ease of reference.

Sulphate of alumina

Sulphate of alumina ($Al_2(SO_4)_3.18H_2O$) is also known as aluminium sulphate or (incorrectly) as alum. It is a very commonly used co-agulant and can be bought in various forms—aluminoferric (in blocks), granulated, kibbled (in lumps) or liquid. Where 10 kg of this salt contains the equivalent of 1.53 kg of aluminium oxide it can theoretically react to produce 2.34 kg of floc. It is sold on the basis of its Al_2O_3 content and is graded (and priced) as 14–15%, 16% or 17–18%.

Theoretically 1 mg/l of alum will react with 0.45 mg/l of natural alkalinity expressed as $CaCO_3$, 0.28 mg/l lime as CaO (quicklime), 0.35 mg/l lime as $Ca(OH)_2$ (hydrated lime) or 0.48 mg/l as Na_2CO_3 (soda ash). Its reaction when added to water is with the natural or added alkalinity. It is readily soluble. Its pH range is 5.5–8 for clarification and 9–10.5 for softening. The solution is corrosive and needs to be stored in tanks with a corrosion-resistant lining.

In all cases aluminium hydroxide $Al(OH)_3$ (floc) is formed according to the alkalinity present. For the natural case:

$$Al_2(SO_4)_3 + 3Ca(HCO_3)_2 \rightleftharpoons 2Al(OH)_3 + 3CaSO_4 + 6CO_2$$

With lime added:

$$Al_2(SO_4)_3 + 3Ca(OH)_2 \rightleftharpoons 2Al(OH)_3 + 3CaSO_4$$

With soda ash added:

$$Al_2(SO_4)_3 + 3Na_2CO_3 + 3H_2O$$
$$\rightleftharpoons 2Al(OH)_3 + 3Na_2SO_4 + 3CO_2$$

In the above equations, and all quoted hereafter, insoluble products (precipitates) are in italics.

The minimum dose is 5 mg/l but the average dose commonly encountered in practice lies in the range 15–50 mg/l. If the water is very turbid the dose might be even more but there is no straight line correlation. Results can be obtained only by jar tests. The lime dose required by soft waters with insufficient alkalinity varies with the alkali used and must again be determined by jar test; it is commonly about 40% of the alum dose.

Sodium aluminate

Sodium aluminate is a compound of sodium oxide and aluminium oxide. It is a white powder almost invariably used in conjunction with sulphate of alumina. The two are never mixed before dosing, the sodium aluminate always being put in separately about 30 s before the alum; the dose used is normally 5–10% of that of the alum. Sodium aluminate is mainly useful in alkaline waters, reacting with the natural alkalinity to give insoluble and flocculant calcium aluminate: clarification is better, the floc is denser, and coagulation is effective even at low temperatures over a wider pH range. It is widely used in lime or lime–soda softening plants in doses of about 15 mg/l. Curiously enough it is not universally effective, no beneficial effect being apparent in certain waters, and therefore it should be tried on an experimental scale before adoption. On the alkaline waters of the Tigris it was ineffective, whereas on similar waters in the Midlands of England it has been used with success.

Iron salts

Certain iron salts can be used as coagulants and when available are normally cheaper, produce heavier flocs and operate over a wider pH range than sulphate of alumina. However, they normally require to be used with lime and are a little more difficult to control. In this connection the following problems might be listed:

(a) special materials may have to be used to line the storage containers;

(b) iron salts tend to cake in humid locations;

(c) iron salts are dirty to handle, causing staining;

(d) sludge is more difficult to dispose of without giving rise to complaint;

(e) lime must generally be added, usually in advance;

(f) while often superior to alum in end results, bigger flocculation chambers may be required.

Ferrous sulphate. Copperas ($FeSO_4.7H_2O$) contains 29% of Fe_2O_3. It is supplied in small lumps or as light-green crystals. It is recommended for use on alkaline bicarbonate waters, and should be accompanied by sufficient lime or caustic soda to raise the pH to 8.5–10. It is not widely used on domestic supplies but is encountered in industrial softening plants and in the treatment of trade wastes. It cannot be used on soft, coloured or turbid waters where floc formation occurs at a pH of about 6, as the ferric hydroxide forms from the reaction of ferrous hydroxide only in the presence of lime at high pH values. The reactions are

$$Ca(HCO_3)_2 + FeSO_4 = Fe(OH)_2 + CaSO_4 + 2CO_2$$
$$2Fe(OH)_2 + O + H_2O = 2Fe(OH)_3$$

It is hygroscopic and tends to clog in dry-feeding equipment and should therefore be added as a solution.

Chlorinated copperas. Theoretically 8 parts of copperas are oxidized by 1 part of chlorine to form a mixture of ferric chloride and ferric sulphate, both of which precipitate. In practice a little more chlorine is added. Chlorinated copperas is extremely effective over a wide range of pH. It is particularly effective in cold water and when manganese is present, both conditions normally presenting problems. It also fits in very conveniently where there is a need for pre-chlorination.

$$6FeSO_4 + 3Cl_2 = 2Fe_2(SO_4)_3 + 2FeCl_3$$
$$2FeSO_4 + 3Ca(HCO_3)_2 + Cl_2 = 2Fe(OH)_3 + 2CaSO_4 + CaCl_2 + 6CO_2$$

The ferric hydroxide formed is positively charged and acts in the same way as aluminium hydroxide in neutralizing the negative charges on colloidal particles.

Ferric chloride is extremely corrosive and difficult to handle, store and apply. It is available as $FeCl_3$ in the anhydrous, lump or crystal, or liquid form. Because of the handling problem it presents it is not very common but is used at Tehran with excellent results at doses varying between 4 mg/l and 120 mg/l according to the widely ranging degree of turbidity. It is actually made at the waterworks in Tehran. The reactions are

$$2FeCl_3 + 3Ca(HCO_3)_2 = 2Fe(OH)_3 + 3CaCl_2 + 6CO_2$$
or
$$2FeCl_3 + 3Ca(OH)_2 = 2Fe(OH)_3 + 3CaCl_2$$

Obviously the alkalinity has to be increased at the higher rates of dosage.

Ferric sulphate is an expensive product which is difficult to dissolve but it has advantages in certain processes such as decolorization of waters of low pH, removal of manganese at high pH, clarification of water of low temporary but high permanent hardness, clarification of lime-softened water, operation at very low temperatures, and reduction of silica in boiler feeds. The reactions are

$$Fe_2(SO_4)_3 + 3Ca(HCO_3)_2 = 2Fe(OH)_3 + 3CaSO_4 + 6CO_2$$
$$Fe_2(SO_4)_3 + 3Ca(OH)_2 = 2Fe(OH)_3 + 3CaSO_4$$

COAGULANT AIDS

Weighters

Certain relatively clear waters are sometimes difficult to settle. This can sometimes be rectified by the addition of weighter or artificial sediment. Various forms of clay such as bentonite or fuller's earth have been used in doses of 10–100 mg/l as an aid to coagulation with sulphate of alumina. This is a fairly rare technique, having proved rather expensive in many cases.

Lime

Lime is widely used with other coagulants throughout the world for the various purposes described above. It can be bought as CaO (quicklime) or $Ca(OH)_2$ (slaked or hydrated lime). Locally available lime can normally be bought in most countries but varies considerably in constitution and purity. Quicklime is stronger and cheaper than hydrated lime but it is troublesome to treat and handle, and is normally used when hydrated lime is not readily available. Lime is normally fed into the process as a suspension of lime in supersaturated lime water but it tends to clog. Constant attention to the pumps and feed lines is necessary to maintain correct dosage. An admixture of Calgon (sodium hexametaphosphate) helps to alleviate this problem. Quicklime is slaked by adding about 4 times the quantity (by weight) of water to each portion of lime. Heat is generated and the slaking action takes about 90 min.

Sodium carbonate

Sodium carbonate (Na_2CO_3), known as soda ash, can be used to increase alkalinity in place of lime. It is more expensive but easier to handle and apply and therefore is occasionally preferred. However, its main use is in conjunction with lime for removal of permanent hardness. When applied as a solution it is generally mixed with at least 12 times its own weight of water.

Activated silica

Silica sol is highly effective as a coagulant aid. It is prepared from sodium silicate 'activated' by various chemicals including chlorine, sodium bicarbonate, sulphuric or hydrochloric acid and carbon dioxide. When used with a normal coagulant in very small doses (of the order of 1–2 mg/l) it reduces the amount of coagulant needed and increases the density of the floc. It is batch-mixed and has to be allowed to 'age' before being put into the water. It is never mixed with the other coagulants: all are introduced separately, lime generally first, but the exact order is best decided by experiment. Curiously enough, the order varies with different waters and seems to matter in each case.

In a typical example, 100 kg of sodium bicarbonate was added to four 1 m³ tanks of water. In a fifth tank of water 280 kg P84 grade of sodium silicate was added. The two batches were mixed and left to age for 30 min, at which time a distinct blue haze developed. Then a final dilution to 21.4 m³ was made and the batch put into use. The concentration of silica in that particular solution was about 1.5%.

Used on a soft moorland water from the Burnhope Reservoir in the north of England (cf. Appendix 1) activated silica materially lengthened the periods of filter operation between washes and improved the quality of the filtrate. It was noted that better results were obtained if the period of time between injection of the alum and injection of the activated silica was increased. This was easy to do because both chemicals were being injected into a long inlet main, and the point of injection of the alum was moved upstream about 130 m. Even earlier injection was experimented with but no additional benefit was attained and it would appear that once thorough mixing was being affected no further improvement could occur. The

Table 4.1. Preparation of activated silica: quantity of activating agent required to activate silica produced as SiO_2 from sodium silicate having a mean weight ratio $SiO_2:Na_2O$ of 3.3:1 (by courtesy of Wallace and Tiernan Ltd)

Activating agent	Formula	Ratio agent/silica
Chlorine	Cl_2	0.5:1
Hydrochloric acid	HCl	0.33:1
Sulphuric acid	H_2SO_4	0.4:1
Sodium bicarbonate	$NaHCO_3$	0.35:1
Aluminium sulphate	$Al(SO_4)_3.18H_2O$	0.8:1

optimum dose of activated silica was 1.5–2 mg/l, lower doses being less effective and higher doses giving no improvement.

Table 4.1 shows the quantity of activating agent required to activate silica.

In some areas the cost of activated silica is about twice the cost of sulphate of alumina, and it therefore follows that if every part of activated silica reduces the alum dose by at least two parts, an overall cash saving should result. This ratio can be adjusted to suit local differentials in the prices of chemicals.

Polyelectrolytes

There are a number of coagulant aids on the market under various brand names. Most of these materials are synthetic, organic, water-soluble, high-molecular-weight polymers. Known as polyelectrolytes, a term which covers a wide variety of compounds, they are often starch-based or polysaccharides. Although their use on potable water has not yet been universally approved (they are banned in France, for instance), their use is extending in most sophisticated countries. Their high cost is claimed to be offset by the very small dosage (of the order of 0.1 mg/l) and the saving in coagulants, but unless dosage is very carefully controlled their employment may be counter-productive. Because of the control required it is somewhat doubtful whether they can yet be recommended for use except where experienced waterworks chemists are available.

As they are synthetic, their composition and molecular size can be varied to suit operational requirements. Polyacrylamide, a non-toxic linear resin, can be made electropositive, negative or neutral. In waterworks conditions where alum is being used as the primary coagulant, polyelectrolytes provide a powerful auxiliary bridging and linking action to promote more rapid settlement. It is normally advantageous to put them into the water after the alum (i.e., after flocculation has commenced).

Easier treatment of sludge is claimed from the use of polyelectrolytes, but is still not proven.

The necessity to dose at the precise rate required was noted at the Hampton Loade works of the South Staffordshire Water Company where the optimum amount of LT 22 was 0.008 mg/l at all levels of raw water turbidity, and variations upwards or downwards led to a sharp fall in effluent quality.

The same is true where polyelectrolytes are used after the basins as filter aids. In this case the basins produce a microfloc on which the polyacrylamides work. The governing rule appears to be that the dose should not be so high as to cause too rapid a rise in head loss and not so low as to permit premature breakthrough.

WATER SOFTENING AND REMOVAL OF IRON AND MANGANESE

Water softening and removal of iron and manganese are covered in Chapter 12. This section concentrates on the chemical reactions of the precipitation methods, and removal of the precipitates.

Water softening

Temporary hardness is caused by the presence of the bicarbonates of calcium, magnesium and sodium. If the water is heated, CO_2 is driven off, leaving insoluble carbonates (i.e., the familiar white scale found in kettles and hot water systems). These carbonates can be formed and settled out in a treatment plant by dosing with lime. This is often referred to as Clark's process. The reactions are as follows (the insoluble precipitates are in italics):

$$Ca(HCO_3)_2 + Ca(OH)_2 = 2CaCO_3 + 2H_2O$$
$$Mg(HCO_3)_2 + 2Ca(OH)_2 = Mg(OH)_2 + 2CaCO_3 + 2H_2O$$
$$2NaHCO_3 + Ca(OH)_2 = CaCO_3 + Na_2CO_3 + 2H_2O$$

Permanent hardness is caused by calcium and/or magnesium sulphate or chloride. The calcium sulphate or chloride can be removed by dosing with soda ash to form insoluble carbonates:

$$CaSO_4 + Na_2CO_3 = CaCO_3 + Na_2SO_4$$
$$CaCl_2 + Na_2CO_3 = CaCO_3 + 2NaCl$$

To remove the sulphates and chlorides of magnesium both lime and soda must be added. The lime–soda process is as follows:

$$MgSO_4 + Ca(OH)_2 + Na_2CO_3 = Mg(OH)_2 + CaCO_3 + Na_2SO_4$$
$$MgCl_2 + Ca(OH)_2 + Na_2CO_3 = Mg(OH)_2 + CaCO_3 + 2NaCl$$

The calcium and magnesium salts that cause hardness are removed by precipitation: those left in do not cause hardness.

Iron and manganese removal

Removal of iron and manganese is by precipitation and filtration, the precise form of chemical treatment varying widely to suit the state of combination of the elements.

When iron is present in water it exists in the soluble, ferrous state, often as ferrous bicarbonate ($FeH_2(CO_3)_2$). When oxidized by chlorine or aeration it turns into an insoluble ferric form $Fe(OH)_3$, which can be removed by settlement and filtration.

Manganese is more difficult to remove than iron and is even less tolerable. However, it can be turned into a precipitate at a reasonably high pH value.

Design considerations

The precipitates formed in softening and in the removal of iron and manganese are heavy and settle quickly. Very high upward flow rates can therefore be adopted (about 4 m/h). The use of upward-flow basins is fairly universal in softening plants.

Although the precipitates are formed by chemical reactions it is common practice to add standard coagulants to the settling basins to assist in the sedimentation of the fines. Coagulant aids are seldom required, however.

As a lot of sludge is created, care must be taken to adopt basins from which this can easily be removed. Some of the best basins for precipitating solids present certain cleaning problems and the Pulsator, for instance, is rarely found on softening or iron-removal plants. Settling basins on iron-removal and softening plants provide classic examples where ease of cleaning the basin is the paramount consideration, the actual precipitation of the solids presenting little difficulty.

5: Mixing and flocculation

When coagulants are added to water and thoroughly mixed, the reaction is almost instantaneous. As soon as floc forms, a further gentle stirring is advantageous in order that the floc particles may coalesce and grow bigger. There is a similarity between the two actions in so far as the water is stirred. However, the first action, preceding floc formation, must be violent; the second, following floc formation, should be gentle.

Both these actions occur naturally to some extent in any basin. If coagulants are fed into an inlet pipe or channel, some mixing is bound to take place due to turbulence. When the dosed water carrying floc finally passes into the settling basin through inlet ports a certain rolling motion is inevitable, which can be accentuated by baffles in a horizontal-flow basin, or in an upward-flow basin by the sludge blanket. There are therefore many existing basins working quite well without special mixing or flocculation chambers because the movement of the incoming water provides naturally for a certain amount of action to take place.

However, in many basins there is ample evidence that better results can be obtained if mixing and flocculation are intensified. In recent years much research has been done on both types of action and sound theoretical rules have been laid down which can be found in standard text books; that by Fair et al.[1] is recommended, and is quoted extensively throughout this chapter.

There are two methods of approach: the mixing and flocculation can be carried out either by mechanical means in specially built chambers, or in a suitably baffled channel or interconnected chambers. The latter method requires no mechanical equipment but lacks flexibility because the system can be designed for maximum efficiency only at one rate of flow and at one temperature, whereas the speed of mechanical paddles can be adjusted to suit the variations of flow, temperature and silt conditions. However, the cost and added complexity of mechanical equipment introduce additional complications, to be avoided in a developing country, and in practice a sinuous inlet channel preceded by violent mixing generally provides a reasonably effective solution.

MIXING

If the pipe or channel through which the incoming water enters the basin is so dimensioned as to ensure a velocity >1 m/s and if the channel is directed at an end wall so that the flow is forced to reverse abruptly through $180°$, any coagulants introduced into the water before that sudden reversal will be adequately mixed and floc will form almost instantaneously. Adequate turbulence can also be induced by passing the water through a measuring flume in which a standing wave is formed, or by stirring either with high speed paddles, or hydraulically by pump-induced turbulence in a small chamber. Good results have been observed where the water is made to fall freely over a weir. Normal flow along a typical entry channel does not seem to be enough.

Adequate mixing is absolutely vital to the success of basin performance. As soon as the chemicals have been put into the incoming water a violent swirling action should be induced. With all due respects to the mathematical approach, it is largely a matter of common sense. If the mixing looks violent it is almost certainly good enough.

Where hydraulically agitated flash mixers are used the chamber may have a retention capacity of only about 20–60 s. The water is kept in a state of agitation by a pump of about 0.25 hp per 1000 m³/d throughput. The pump circulates water in the basin by sucking from and discharging into the same chamber. Water falling over a weir with a 30 cm overfall provides much the same effect.

If mixing chambers are used, benefit can be obtained by having more than one and making the water pass through them all. Where this is done, inadequate mixing and short-circuiting will be eliminated.

FLOCCULATION

Following the introduction of coagulants, accompanied by violent mixing, floc forms almost instantaneously but it may be fine and will probably require time to coalesce with neighbouring particles to form bigger masses. This action can be accelerated by gentle stirring. This can be demonstrated in the standard stirring device (Fig. 14.1) which should be found in any waterworks laboratory. In this stirring device, optimum doses of coagulants can be determined under varying pH conditions, and the maximum amount of floc obtainable can be ascertained by direct observation.

Unlike the absolute necessity of thorough mixing, the need for flocculation as a separate process may or may not be essential. Much depends on the nature of the suspended solids. For rivers carrying a coarse and heavy sediment the main problem is to prevent the silt settling and blocking the inlet channels before it reaches the basins.

Flocculation chambers are rarely found on hopper-bottomed upward-flow basins which utilize the sludge blanket effect (although on solids-recirculation tanks, which also operate on the upward-flow principle, they are provided as part of the process). Specially designed flocculation chambers may therefore be essential only before shallow-depth settlers, and for traditional horizontal- or spiral-flow basins where the raw water is carrying a high colloidal content.

Shallow-depth settling may present operational difficulties in developing countries, so separate flocculators are most commonly found before conventional horizontal-flow basins where colloids are a problem, and the complication of additional machinery may be avoided by having the flocculating action imparted to the water by the gently rolling motion resulting from passing the water along a sinuous inlet channel (Fig. 5.1). In practice a channel providing a velocity of flow of about 0.3 m/s, with cross-walls ensuring 12–20 changes of direction through 180° (with well rounded corners), has often proved to be very effective. The emphasis must be placed on the comparative smoothness of flow required. Under no circumstances should velocity or turbulence be such as to break up the floc.

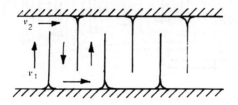

Fig. 5.1. Sinuous flocculation channel

The nominal loss of head in such a channel is 0.5–1.5 m. The same principle may be applied to special flocculating chambers (Fig. 5.2).

Where flocculating chambers are provided with paddles to induce the rolling motion (Fig. 5.3), typical examples indicate basins of 20–45 min nominal retention capacity. As a matter of structural convenience, the basins are generally built with the same width and depth as the settling basins they precede. This gives them the appearance of being preliminary chambers of the main basins. The paddles may revolve on vertical or horizontal shafts, or oscillate. There is sometimes a series of about four paddles, each one capable of being operated at variable speeds to induce a rolling motion.

The total area of the paddles is about 20% of the cross-sectional

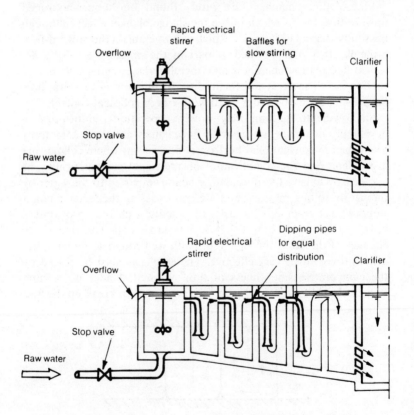

Fig. 5.2. Hydrodynamic flocculators (by courtesy of SA CTE)

area of the chamber and their maximum tip speed should not exceed 0.6 m/s, and be variable downwards to as little as 0.15 m/s. The total horsepower of the driving motors is generally about 0.1 hp for 1000 m³/day, but if the paddles are of the oscillating type the motors have to be a little bigger to overcome a 'stalling point' as the paddles reverse.

THEORETICAL APPROACH

The stirring of water creates differences of velocity and therefore velocity gradients. The average temporal mean velocity gradient in a shearing fluid is denoted by G. For baffled basins or sinuous channels

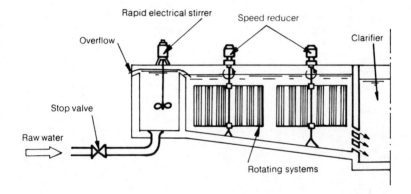

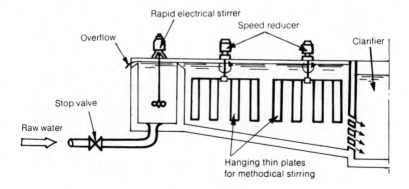

Fig. 5.3. Mechanical flocculators (by courtesy of SA CTE)

$$G = \left\{ \frac{\rho g h}{\eta T} \right\}^{1/2} \tag{5.1}$$

where

G is the velocity gradient, s^{-1}

ρg is the weight of water per unit volume (9.80×10^3 N/m³ at 10°C; 9.79×10^3 N/m³ at 20°C)

h is the head loss due to friction, m

η is the dynamic viscosity (1.31×10^{-3} Pa s at 10°C; 1.01×10^{-3} Pa s at 20°C)

T is the detention time, s.

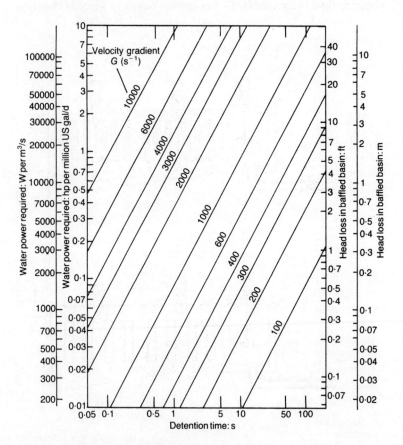

Fig. 5.4. Power required for rapid mixing, 4°C

For mechanical agitation

$$G = \left\{ \frac{P}{\eta V} \right\}^{1/2} \qquad (5.2)$$

where

P is the power consumption, W
V is the volume of fluid, m^3.

The basic equations relating to mixing and flocculation were published a few years ago in the USA by Camp and Stein[2] and more recently presented by Hudson[3] in his book *Water clarification processes*, from which Figs 5.4 and 5.5 are taken. The figures are extensions of each other, both being based on the formula $G = 425(60\text{WHP}/T)^{1/2}$, where WHP = water horsepower per million US

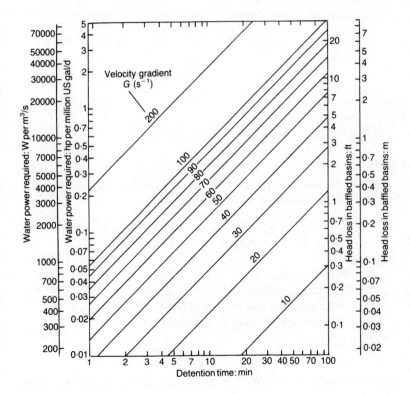

Fig. 5.5. Power required for flocculation, 4°C

gallons per day. This converts to $G = 25.2[(P/Q/T)]^{1/2}$, where P/Q = water power in watts per cubic metre per second; or $G = 25.2(P/V)^{1/2}$.

The total number of particle collisions is proportional to GT, and is greater when there is a degree of turbulence as opposed to general rotation. It has been observed that in many of the more successful mixing basins which are mechanically stirred the GT values given in Table 5.1 apply. Increase in contact time above 120 s achieves little, and excessive GT values can be harmful.

The optimum GT values for flocculation have often been found in the range 30 000–60 000, and there are many basins where $G = 20–50 \ s^{-1}$ and detention time is 30 minutes or so.

Table 5.1. *Recommended GT values for flash mixers*

Contact time T, s	Velocity gradient G, s^{-1}	GT
20	1000	20 000
30	900	27 000
40	750	30 000

Sinuous channels

The permissible loading on a sinuous channel at any given value of GT is (from equation (5.1))

$$\frac{Q}{V} = \left\{ \frac{Q\rho gh}{\eta V} \right\}^{1/2} / GT = \left\{ \frac{Qgh}{\nu V} \right\}^{1/2} / GT$$

where

  Q is the rate of flow, m^3/s
  V is the channel volume, m^3
  g is the acceleration due to gravity (9.81 m/s^2)
  ν is the kinematic viscosity of the water, m^2/s.

The useful power input is

$$P = Q\rho gh$$

where P is the power, W. For each metre of lost head, the useful power input is 9.8×10^3 W per m^3/s. In practice, head losses are commonly 0.15–0.6 m, velocities 0.15–0.5 m/s and detention times 10–60 min.

For a channel with baffles (over and under or round the end), the extra head loss in the channel (in addition to normal channel friction) is

$$h = \frac{n v_1^2 + (n-1) v_2^2}{2g}$$

where h is the additional head loss, m

$n-1$ is the number of baffles

v_1 is the velocity between the baffles, m/s

v_2 is the velocity at the baffle slots, m/s.

To approximate, the total channel friction can be obtained by calculation using a Hazen Williams discharge coefficient of $C = 50$.

Example. A sinuous channel has 15 round-the-end cross-walls. Water is passed along with a velocity of 0.2 m/s between the cross-walls and 0.5 m/s round the ends. The flow is 0.3 m³/s and the nominal retention time is 25 min. For a temperature of 10°C, estimate the additional loss of head, the power dissipated, and the G and GT values.

Additional head loss = $(16 \times 0.2^2 + 15 \times 0.5^2)/2 \times 9.81$ m = 0.224 m

$P = 0.3 \times 9.80 \times 10^3 \times 0.224$ W = 659 W

$T = 25 \times 60$ s = 1500 s

$V = 0.3 \times 1500$ m³ = 450 m³

$G = [659/(1.31 \times 10^{-3} \times 450)]^{1/2}$ s^{-1} = 33.4 s^{-1}

$GT = 33.4 \times 1500 = 5.01 \times 10^4$

Mechanical flocculation

In considering the design of a powered, paddle-operated flocculation basin the following is a method of calculation based on the formula

$$P = \tfrac{1}{2} C_D \rho A v^3 \tag{5.3}$$

where

P is the power, W

C_D is the coefficient of drag of the paddle (1.8)

ρ is the mass of water per unit volume (1.000×10^3 kg/m^3 at 10°C, 0.998×10^3 kg/m^3 at 20°C)
A is the area of the paddle, m^2
v is the relative velocity between paddle and water, m/s.

Example (*based on an example given by Fair et al.*[1]). A flocculator designed to treat 75 000 m^3/day is 30 m long, 12 m wide and 4.5 m deep. It is equipped with 0.3 m paddles supported parallel to and moved by four horizontal shafts which rotate at a speed of 2.5 rev/min. The centre line of the paddles is 1.7 m from the shaft, which is at mid-depth of the tank. Two paddles are mounted on each shaft, one opposite the other. Assuming that the mean velocity of the water is approximately one quarter the velocity of the paddles, that the drag coefficient of the paddles is 1.8, and that the water temperature is 10°C, find

(a) the velocity differential between the paddles and the water;
(b) the useful power input and the energy consumption;
(c) the detention time;
(d) the values of G and GT.

The paddle velocity is $2\pi \times 1.7 \times 2.5/60$ m/s = 0.445 m/s. The velocity differential is $v = (1-0.25)0.445$ m/s = 0.334 m/s.

Because the area of the paddles is $A = 12 \times 2 \times 4 \times 0.3$ m^2 = 28.8 m^2, and the coefficient of drag $C_D = 1.8$, the useful power input, by equation (5.3), is $P = 0.5 \times 1.8 \times 1.000 \times 10^3 \times 28.8 \times 0.334^3$ W = 966 W. The net energy consumption per unit volume treated is therefore $966 \times 24 \times 3600/75\,000$ J/m^3 = 1.11 kJ/m^3. For electrical drive there must be added the energy required to overcome mechanical friction and to provide for electrical losses in the lines and motor. (In practice, flocculators consume about 6 kJ/m^3 (i.e., 2 kW h per 1000 m^3.)

As the volume of the tank is $30 \times 12 \times 4.5$ m^3 = 1620 m^3, the detention period $T = 1620 \times 24 \times 60/75\,000$ min = 31.1 min.

By equation (5.2), $G = (964/1.31 \times 10^{-3} \times 1620)^{1/2}$ s^{-1} = 21.3 s^{-1}. $GT = 21.3 \times 31.1 \times 60 = 3.97 \times 10^4$.

6: Theory and principles of sedimentation

When silt-laden water is admitted to the still conditions of a sedimentation basin its velocity tends to fall to zero, its capacity to transport solids disappears, and the solids begin to settle.

It has long been established that a discrete particle* settling freely through water quickly attains a constant velocity[1]

$$v_s = [(2g/C)\,(s-1)\,V/A_c]^{1/2} \qquad (6.1)$$

where v_s = velocity of settlement, cm/s

g = acceleration due to gravity (981 cm/s²)

C = drag coefficient*

s = specific gravity of the particle

V = volume of the particle, cm³

A_c = projected area of the particle, cm².

It can be seen that increases in the size of the particle (V) and the drag coefficient (C) speed up and slow down v_s respectively.

Most of the spherical particles of concern in water treatment settle in accordance with a modified form of the above formula known as Stokes' law in which v_s can be written[1]

$$v_s = \frac{g}{18}\,(\rho_1 - \rho)\,\frac{d^2}{\eta} \qquad (6.2)$$

where ρ_1 = density of the particle, g/cm³

ρ = density of the fluid, g/cm³

η = dynamic viscosity* of the fluid, g/(cm s)

d = diameter of the particle, cm.

---

* See Glossary.

In terms of the kinematic viscosity v (cm^2/s) the above formula can be written

$$v_s = \frac{g}{18} \frac{(\rho_1 - \rho)}{\rho} \frac{d^2}{v}$$

(The kinematic viscosity is the dynamic viscosity divided by the density.)

For settling in the Stokes' law region, the drag coefficient C in equation (6.1) is 24/(Re) (where (Re) is the Reynolds number*), and thus it decreases as the Reynolds number increases (Fig. 6.1).

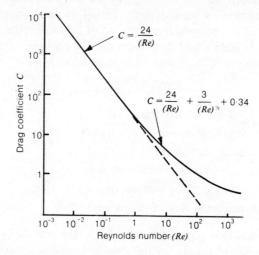

Fig. 6.1. Newton's C for a range of Reynolds numbers[1]

Reynolds number is inversely related to kinematic viscosity, which decreases with rising temperature (Table 6.1). It therefore follows that higher water temperatures decrease the drag coefficient and thus increase the rate of settlement. Similar particles take twice as long to settle in a Canadian winter than in a tropical summer and the basins should be sized accordingly.

It is a matter of everyday observation that larger particles settle more quickly than smaller ones. Figure 6.2 shows settling velocity plotted against particle diameter.

---

* See Glossary.

Table 6.1. Variation of viscosity with temperature[2]

Temperature, °C	Density of water, g/cm³	Dynamic viscosity η, 10^{-2} g/(cm s)	Kinematic viscosity ν, 10^{-2} cm²/s
0	0.99987	1.7921	1.7923
4	1.00000	1.5676	1.5676
10	0.99937	1.3097	1.3101
20	0.99823	1.0087	1.0105
30	0.99568	0.8004	0.8039

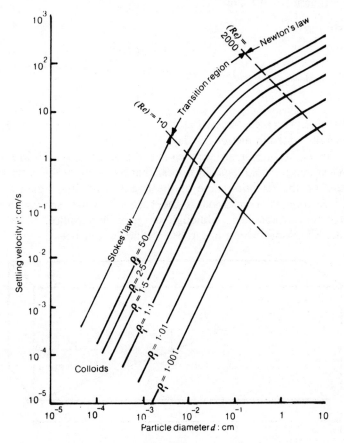

Fig. 6.2. Settling velocities of discrete particles in quiescent water at 10°C[2]

SETTLEMENT IN HORIZONTAL-FLOW BASINS

From this theory of simple, unaided settlement it follows that in a rectangular horizontal-flow basin as portrayed in Fig. 6.3, a particle settling at v_s cm/s and being carried horizontally by water flowing at velocity v cm/s would follow the inclined path AB, and by comparing similar triangles the particle would just reach the bottom when $v_s/v = D/L$. If

Q is rate of flow, m³/s
W is width of basin, m
A is area ($= WL$), m²

then

$$Q = (v/100)\ WD$$

and

$$v_s = 100\ Q/A.$$

The quantity Q/A is known as the overflow rate or velocity. It is generally stated in metres per day.

The importance of surface area in the simple theory of settlement is clear and it would logically appear that the depth of the basin and therefore the retention time of the liquid is of little significance. This theoretical conclusion is the basis on which shallow-depth settlement is founded. Although of great importance, there are practical reasons why it has to be modified in practice.

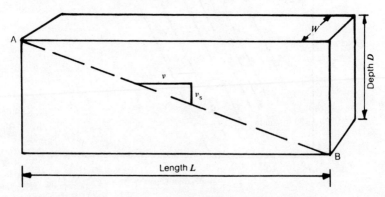

Fig. 6.3. Theoretical settlement in a horizontal-flow basin

Laboratory experiments

Much research has been done by Camp, Imhoff and others[1,3] on the settling velocities of discrete particles and typical figures for v_s are given in Table 6.2. They are of general value but are difficult to apply directly. The figures are of somewhat academic importance because the particle sizes in any given suspension are very variable, and the particles rarely remain discrete. It is usual to obtain settling velocity analyses by laboratory tests.

Under typical waterworks conditions, with flocculation taking place, the simplifying assumption relating to discrete particles does not apply and it is easy to demonstrate that retention time is also of importance, and therefore settlement is not independent of basin depth. One such experiment makes use of a cylinder, equal in depth to the proposed basin, with draw-off points at different levels, as shown in Fig. 6.4. The turbidity of samples from each of the draw-off points A, B, C and D can be measured at increasing time intervals as the water clears slowly from the top. The practical importance of this test is that in a horizontal-flow basin the average turbidity at the outlet weir is that of the entire vertical cross-section of the basin and the effectiveness of the basin after any given time is reflected by the average of the turbidity readings at all the outlets.

Table 6.2. Settling velocities of particles in water

Diameter of particle, mm	Settling velocity, cm/s			
	Sand, specific gravity 2.65		Alum floc,[4] specific gravity 1.05*	
	Temperature 10°C	Temperature 20°C	Temperature 10°C	Temperature 20°C
1	11	12	0.07	0.1
0.5	6	7.8	—	—
0.1	0.8	1.04	—	—
0.05	0.15	0.2	—	—
0.01	0.01	0.013	—	—
0.005	0.0015	0.0016	—	—

* Calcium softening precipitates with a specific gravity of 1.2 settle about 3 times as fast as alum floc.

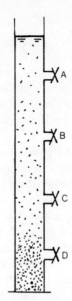

Fig. 6.4. Settlement rate experiment

In a widely quoted experiment carried out by Camp[1] the results in Table 6.3 were noted. They show that removal is not independent of depth and is influenced by both overflow rate and retention time. For elapsed times of 20 min and more, some of the silt is assumed to have reached the floor of the tube, from where it will not be picked up.

Table 6.3. Results of settlement experiment[1]

Elapsed time, min		0	20	40	80	160	300
Concentration ratio	Sampling point A	1.0	0.82	0.66	0.47	0.28	0.21
	B	1.0	0.83	0.69	0.50	0.33	0.25
	C	1.0	0.85	0.72	0.54	0.38	0.30
	D	1.0	0.87	0.75	0.59	0.43	0.36
	Average	1.0	0.843	0.705	0.525	0.355	0.280
Removal ratio		0.0	0.157	0.295	0.475	0.645	0.720

In this experiment it can incidentally be seen that if the basin were shallower, the degree of clarification in a given time would be better. For instance, after 160 min retention the average concentration ratio of silt from points A and B would be 0.305, as opposed to 0.355 for a basin of full depth. The removal ratio in the shallower basin would be 0.695 as compared with 0.645, and the logical conclusion is that a shallower basin is as good as or better than a deep one.

This experiment therefore supports the previous mathematical demonstration that a shallow basin compares well with a basin of greater depth, if Q/A is the same in both cases.

SETTLEMENT IN UPWARD-FLOW TANKS

Settlement in the upper regions of an upward-flow basin is controlled by making the area (A) of the tank sufficiently big so that $v < v_s$, where v is the upward velocity of the water ($= Q/A$) and v_s is the settling velocity of the particle. When this condition is achieved the particle must be settling through the rising water and clarification must result.

In practice the upward velocity of the water is kept down to about half the settling velocity of the floc particles. The normal velocity of settlement of well formed floc is about 3 m/h.[4] If coagulant aids are used this may become 6–10 m/h, and where the floc is consolidated by pulsing or other devices it may be somewhat more. In a softening plant the settling velocity of the particles of calcium carbonate is about 8 m/h. Applying the commonly used factor of 0.5, many upward-flow basins are provided with surface areas designed to limit the upward velocity of the water to the values shown in Table 6.4.

Table 6.4. Design upward water velocities for upward-flow tanks

	Velocity, m/h
River supply, normally coagulated	1.5
River supply, with coagulant aids	3–5
River supply, with coagulant aids and floc improvement by pulsing	6
Softening plants	4.2

As in the case of the horizontal-flow basins it would appear in theory that the depth of the tank is not particularly significant and that surface area is the only important factor. However, as demonstrated in Table 6.3, in the case of all conventional basins this conclusion is not correct. Depth is important for a number of practical reasons. This is evident from the fact that the vast majority of existing basins are fairly deep, being 3–6 m in most cases.

THE MERIT OF ADEQUATE DEPTH

Except in the special case of shallow-depth basins (which probably do not account for 1 % of all the basins actually constructed) all basins have appreciable depth (i.e., 3 m or more). Area × depth = volume, and (erroneously or not) it is a well-established custom to relate basin capacity to the hourly rate of throughput and classify basins in terms of 'hours of nominal retention'. A capacity of 3–6 h is typical of many horizontal-flow basins but vertical-flow basins are invariably much smaller.

The reasons for giving adequate depth to basins are as follows.

The theory of settlement is based on the concept of the discrete particle. A discrete particle by definition remains separate and unchanged in volume whereas, in actual practice, silt in a basin, in the presence of coagulants, grows in size and joins up with its neighbours. The bigger the particle becomes the quicker it settles; and the greater the distance through which it can fall the more of its smaller neighbours it adheres to and removes from suspension. Therefore depth (i.e., falling distance) has value.

The theory of settlement pays no regard to the accumulation of silt. Between periodic removal operations it is generally convenient to let it build up on the basin floor, and some surplus capacity is extremely useful to provide storage space.

In a horizontal-flow tank there are four zones: the inlet zone where silt and water travel horizontally; the settling zone in which the silt moves downwards relative to the water; a bottom zone where sludge accumulates, compacts and is stored between removals; and an outlet zone where clarified water and any remaining fines are carried to the outlet weirs. During settlement a particle grows and its rate of descent increases until it reaches the bottom zone where its rate decreases because of the accumulation and resistance of other particles. Thus its path is not as shown in Fig. 6.3.

The theory of settlement assumes gentle non-turbulent flow, but it is clear that, in very shallow horizontal-flow basins, velocities would tend to increase to a point where excessive velocity and turbulence could interfere with the whole settlement process. Because $v = Q/WD$ (WD is the cross-sectional area of the basin), D plays a part in keeping v within reasonable limits (i.e., <30–90 cm/min). Above velocities of this order some of the finer particles would remain in suspension.

In upward-flow tanks the water enters at the bottom and in its passage upwards has to flow through a zone of previously settled sludge. In so doing the newly arriving silt particles adhere to the old sludge and only relatively clear water passes into the upper regions of the tank. This action depends on the presence of sludge which has to be accommodated in fairly deep tanks and is the main reason for the effectiveness of the upward-flow process. Basins in which sludge accretion is promoted are sometimes called 'solids contact' or 'sludge blanket' tanks, and they form a high proportion of all basins now operating.

SHALLOW-DEPTH SEDIMENTATION

Efforts to realize the undoubted advantages inherent in shallow-depth sedimentation have continued to be made. The principle of making settling basins as shallow as possible was first stated by Hazen in 1904. Even in the 1950s Camp was proposing settling basins with depths of only 15 cm and detention times of 10 min. Many early attempts actually to do this were unsuccessful because of the unstable hydraulic conditions in the shallow basins and the difficulty of removing sludge.

In the last few years various techniques have been evolved to overcome both these problems. In the USA there are a number of plants where the water to be settled is passed at a velocity of about 15 cm/min through small-diameter tubes[5] having a large wetted perimeter in relation to their cross-sectional area. Laminar flow results because of the low Reynolds numbers, and as the vertical distance through which the particles have to settle is only a few centimetres, detention times of as little as 10 min have been attained. If the tubes are steeply inclined (at 60° to the horizontal) the silt rolls downward along the invert and out of the bottom onto a silt-removal mechanism of the moving belt type.

The angle of tilt is critical: too little, and sludge clearance is not attained; too much, and the tube acts on the upward-flow principle, all benefit due to shallow-depth horizontal-flow action being lost. In spite of the marked inclination of the tubes it is important to remember that the settling action is governed by horizontal-flow principles which are in no way different from those described in the early part of this chapter.

Basins working on a somewhat similar principle have rows of plates which are arranged in a zig-zag pattern and suspended from hangers (cf. Fig. 8.3). The sloping plate basins originated in Japan but have spread into adjoining countries, notably Korea. These plates are 8 cm apart measured horizontally and the flow of water is completely horizontal and parallel to the plates. The silt has only a very short distance to fall onto the plate below; flow is laminar and the silt slides downwards from plate to plate, finally falling onto a belt scraper on the floor of the tank. The design of these plated basins is covered by Japanese patents.

Thus the problems of turbulent flow and silt removal have been overcome and savings in basin size effected.

However, it would appear that the main sphere of usefulness of shallow-depth settlers will lie in the ease with which tube nests can be inserted into existing basins of conventional design, thus increasing their output capacity without requiring additional land or civil structures. There are many examples which confirm the value of shallow-depth settlers when used this way (cf. Chapter 8).

SPIRAL-FLOW BASINS

The vast majority of settling basins act on either the vertical- or the horizontal-flow principle but there are a few in which the water follows an upward spiral path. These originated in Alexandria, Egypt, to deal with exceptionally difficult water from the Sweetwater Canal, through which a portion of the flow of the Nile finds its way into the delta.

Water is admitted tangentially at the base of the circular tank and exits over a peripheral weir extending for about one third of the circumference at the top of the tank.

The basins have been quite successful, although probably not to the extent originally claimed, but only a few others have been constructed.

Spiractor

The Spiractor is a highly efficient settling tank used in the lime settling process. It depends for its efficiency on the swirling agitation of the softening reaction. It is described in Chapter 8.

DISSOLVED AIR FLOTATION[6]

It can be seen from equation (6.2) that when the separation velocity $(gd^2/18\eta)\,(\rho_1 - \rho)$ has a positive value particles will settle, whereas if it has a negative value particles will rise and float. Certain suspended solids like algae have a tendency to float anyway; if air bubbles attach themselves, even heavier particles can be made to rise. The upward velocity of an air bubble in water varies with its size but is quite fast, and even when the bubbles are supporting particles, upward separation velocities are many times greater than those of particles settling downwards in a conventional basin. Floating 'scum' is also appreciably drier than settled sludge and this is an advantage where post-process treatment of sludge is envisaged.

When this principle was first used, air was injected at the bottom of the basin in the hope that bubbles would cling to the suspended particles and cause them to float, but it was found that this effect was difficult to control and caused turbulence and floc breakage.

A modern patented process uses dissolved air at 5–6 atm which is made to supersaturate some 5–20 % of the settled water which is then reinjected at the basin inlets. With the use of special equipment, microbubbles of air are released which are widely diffused throughout the incoming raw water.

Figure 6.5 shows a typical plant layout. Due to high separation velocities it is claimed that the nominal retention time of the basin can be reduced to 40–80 min, and additionally that, as the size of the floc is of less importance, savings may also be made on coagulants and the size of flocculation chambers and equipment. In comparison with other upward-flow basins, however, there is no sludge blanket effect and this is a major disadvantage, and there is the additional cost of compressers and diffusion equipment. At least one leading manufacturer of both types of equipment appears to favour the sludge blanket tank rather than the alternative working on the flotation principle, and certainly flotation tanks are not yet widely used.

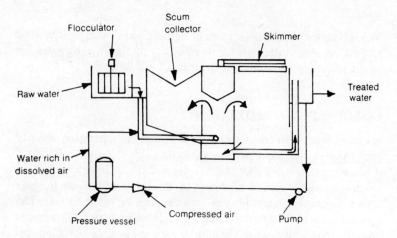

Fig. 6.5. Flotation plant

No patent attaches to the principle of flotation but its success depends on the effectiveness of the means of forming the adherent air bubbles and patents attach to the apparatus for doing this.

7: Settling basins: practical considerations and choice

Pure theory is of very little use in designing a settling basin. For one thing it is difficult to predict the worst conditions under which the basin will have to operate. Although laboratory tests on a series of samples will give an indication of the most suitable types of basin and the required coagulant doses, and of optimum floc formation and settling velocity, generous factors of safety have to be allowed before the results can safely be applied in practice.

COMPARISON OF RETENTION TIMES AND FLOW RATES

In a horizontal-flow basin it is usual to provide for a retention time of 3 times the period required to obtain the desired clarity in still water, as given by experiment in a vertical tube.

In vertical-flow basins it is usual to maintain an upward water velocity of 0.5 times the normal settling velocity of the silt particles. As well formed alum floc settles at about 3 m/h under normal conditions, a typical upward-flow tank (without coagulant aids) is often designed to run with upward flow velocities of 1.5 m/h.

Horizontal-flow basins

In England, where the climate is temperate and the silt loads are below average, horizontal-flow basins vary in size between capacities of 1 h and $3\frac{1}{2}$ h of nominal retention time. Assuming a water depth of about 3.5 m, a '3 h basin' would have a surface overflow rate of about 28 m/day. In many tropical countries, where the rivers tend to be more turbid, basins are commonly of 4 h retention capacity (about 18 m/day surface overflow rate). In really difficult cases where there is heavy turbidity, a high proportion of colloids or low temperatures, the basins may have to have a nominal retention capacity of 6–16 h, but the latter would be most unusual. All the above retention times assume that coagulant aids are not used.

In England, with the exception of the River Severn, the total suspended solids rarely exceed 500 mg/l. On the Severn they might reach 1000 mg/l on occasions. In other countries much more turbid water may be experienced. On the Tigris at Bagdad, suspended solids of 10 000 mg/l occur frequently, and on occasions they can be much higher.

Horizontal-flow basins are at their best under very silty conditions and as a result are becoming less common in England, where the raw water suits upward-flow tanks. The effectiveness of horizontal-flow basins with silty tropical waters is often under-estimated by those whose experience has been gained elsewhere.

Upward-flow basins

Upward-flow basins are of many patterns and are often of proprietary design. The following upward velocities are commonly encountered in the treatment of river waters:

upward flow, with sludge blanket,
    without coagulant aids   1.5 m/h

upward flow, with sludge blanket,
    with coagulant aids   3.0 m/h

upward flow, with sludge blanket, with
    coagulant aids, and incorporating a
    pulsing motion or other blanket
    improvement   up to 5.5 m/h

For lime softening or iron removal, a typical upward velocity is 4.0 m/h.

DIFFICULT CONDITIONS

The term difficult conditions can be applied when one or more of the following conditions is present:

- (*a*) excessive suspended solids;
- (*b*) coincidence of peak output with peak turbidity;
- (*c*) high colloidal content;
- (*d*) low coefficient of fineness;
- (*e*) low temperature;

(*f*) persistent wind;
(*g*) liability to streaming;
(*h*) overturn of water in the basin.

The times and figures mentioned in the following notes normally refer to coagulated, well flocculated water in which coagulant aids have not been used.

Excessive suspended solids

The effectiveness of a basin declines if the incoming water contains excessive suspended solids. The maximum suspended solids that an upward-flow basin can normally take in its stride is about 900 mg/l. Horizontal-flow basins can withstand a lot more than vertical-flow basins. It would be unwise to expose the latter to suspended solids frequently in excess of 1000 mg/l, whereas the former normally cope reasonably well, especially if their size can be effectively increased by slowing down the throughput. The commonly used surface overflow rate of 18 m^3/day per m^2 together with a basin depth of about 3–3.5 m gives a nominal retention time in a horizontal-flow basin of about 4 h. As a preliminary guide this might be varied up or down roughly in proportion to the ratio between the square root of the maximum suspended solids concentration to the square root of 900. For the typical fairly clear UK river (of about 500 mg/l maximum suspended solids) this would give $(500/900)^{1/2} \times 4$ and so permit the use of a basin of 3 h nominal retention capacity, whereas a fairly turbid tropical water would require about $(2000/900)^{1/2} \times 4$ or about 6 h capacity. This rule of thumb has no theoretical basis but has evolved from many practical examples.

Upward-flow basins can be troublesome to operate when suspended solids exceed 1000 mg/l and if water more turbid than this occurs frequently a small horizontal-flow preliminary basin should also be used. Preliminary basins are necessary only on very bad water indeed if horizontal-flow basins are used, but even then they may be desirable to reduce the size of the main basins and to save coagulants.

Coincidence of peak demand with flood

In countries with cold winters and hot summers like Iran and Iraq, the demand curve rises to very high seasonal peaks, often about 50%

or more above average. If this seasonal peak coincides with maximum suspended solids in the river it imposes very arduous conditions because the treatment plant is severely taxed at a time when maximum output has to be maintained. The two peaks more or less coincide on the Nile and on the Indus but at Bagdad on the Tigris the river is normally at its most turbid at times of low or average demand. In the original plant at Bagdad, the horizontal-flow basins of 4 h nominal retention capacity could therefore be worked at lower speed when necessary. Thus during flood the capacity approximated to an effective 6 h of nominal retention and the basins could normally cope fairly well. Running on 33 % overload in summer the nominal retention capacity could be as low as 3 h but no problems were encountered as during the hot season the Tigris has suspended solids below 100 mg/l, for which 3 h retention is more than ample.

Low coefficient of fineness

The coefficient of fineness is the ratio between the dry silt by weight (mg/l) and the turbidity reading (JTUs), both measured under flood conditions. This ratio is rarely the same for different rivers or even for the same river at different times. However, it is commonly found that waters where the coefficient of fineness is above unity settle more readily than those whose coefficient is less than unity. Basin capacity arrived at by considering the maximum suspended solids could be divided by the coefficient of fineness to adjust the nominal retention time. This should be regarded only as a check on the laboratory settling time results, however.

Low temperature

Settlement occurs in accordance with Stokes' law (Chapter 6): the downward velocity of the settling particles is inversely proportional to the viscosity of the water, which in turn is inversely proportional to the temperature. Thus suspended particles sink more slowly in cold water.

The temperature of the River Tigris in summer is 33°C, and that of the Ottawa River at Ottawa in winter is 0°C. The dynamic viscosity of water at 0°C is 1.79×10^{-2} g/(cm s) and at 33°C is 0.76×10^{-2} g/(cm s). The basins at Ottawa are comparable in size with those at Bagdad, although coagulant aids are used in the former and not in the latter, and the amount of silt in the Ottawa River is trifling in comparison with the Tigris. The adverse factor at Ottawa is mainly the low temperature.

Ottawa River water is typical of the river waters of eastern Canada, being soft and highly coloured. The Tigris is typical of the rivers of Iraq and Iran, being moderately hard but prone to heavy silt concentration in time of flood.

Persistent wind

Settling basins situated near the sea often have trouble because a steady sea breeze can set up surface currents and eddies. This action is common in places like Karachi, Alexandria and various Scottish coastal areas. The remedy is to screen the basin or cover it.

Streaming

The phenomenon known as streaming tends to afflict all horizontal-flow basins but is particularly bad on radial-flow basins. The term describes a condition in which the incoming water does not mingle with the main bulk of water in the basin but passes rapidly through from inlet to outlet in a fairly well defined stream. It is very common, and salt tests have shown that some of the water entering a basin of 4 h capacity actually passes over the outlet weir within a few minutes. When this happens the whole theory of settlement is clearly inapplicable.

This short-circuiting is due to currents set up by waters of different densities caused by differences of either temperature or silt content. The 'stream' may be along the surface or along the basin floor according to the relative temperatures of the incoming water and the ambient air and water in the basin. In the same basin the stream may follow different tracks at different seasons.

Claims have been made that judiciously placed baffle walls have cured streaming, but many engineers find that the eddy-forming-capacity of a baffle wall often causes more problems than it solves, and has the further disadvantage of precluding the installation of mechanical scrapers. A basin with two compartments in series is a more effective answer. Some benefit also results from constructing basins which are long in relation to their width. Such basins have a fairly high Froude number ($(Fr) = v^2/Rg$ where v is the mean velocity and R is the mean hydraulic depth) and there is a school of

Table 7.1. Choice of settling basin: dots signify 'merit'

	Standard of performance in normal conditions	Advantageous use of area	Effectiveness with algae	Extent of use	Lack of dependence on large-scale pre-treatment	Freedom from manufacturers' patents	Effectiveness on small scale	Effectiveness on big works	Cost (overall) (may depend on local circumstances)	Freedom from streaming and overturn	Performance with heavily silted waters	Performance with unskilled operators	Ease of cleaning	Capacity to withstand sudden changes in quality	'Do it yourself'	Suitability for iron removal	Suitability for lime softeners
Conventional horizontal flow	:	·	:	:	:	:	:	:	:	:	:	:	:	:	:	:	:
Multi-story horizontal flow	:	:	:	:	:	:	:	:	:	:	:	:	:	:	:	:	:
Radial horizontal flow	:	:	:	:	:	:	:	:	:	:	:	:	:	:	:	:	:
Shallow-depth tube clarifiers	:	:	:	:	:	·	:	:	:	:	:	:	:	:	:	:	:
Shallow-depth plate clarifiers	:	:	:	:	:	:	:	:	:	:	:	:	:	:	:	:	·
Spiral flow	:	:	:	:	:	:	:	:	:	:	·	:	:	:	·	·	·
Spiractors*		:	:	:	:	:	:	:	:	:	:	:	:	:	:	:	:
Hopper-bottom, upward flow	:	:	:	:	:	:	:	:	:	:	:	:	:	:	:	:	:
Upward flow, with mechanically induced sludge accretion	:	:	:	:	:	:	:	:	:	:	:	:	:	:	:	:	:
Upward flow without sludge accretion	:	:	:	:	:	:	:	:	:	:	:	:	:	:	:	:	:
Upward flow, with hydraulically induced sludge accretion, improved by pulsing or	:	:	:	:	:	:	:	:	:	:	:	:	:	:	:	:	:
Gravilectric cones	·	:	:	·	·	·	:	:	:	:	:	:	·	·	·	·	·

* Used only for softening.

thought which associates moderately high Froude numbers with satisfactory basins, because streaming is virtually eliminated.

Streaming is often very marked in radial-flow basins because the ratio of length of flow to width ($= r/\pi r$) is lower than 1, whereas all the better 'rectangular-plan' basins deliberately have a length/width ratio > 1.

Vertical-flow basins suffer from streaming only if badly constructed in that the draw-off weirs are at different levels, thus encouraging unequal flow to different points on the surface.

Overturn

In hot countries, vertical-flow basins have been known to suffer from daily overturn. Basins of the hopper-bottomed type seem to be particularly prone to this phenomenon, which in countries like Malaysia and Sarawak has been observed in the early afternoon on certain works. The water in the basin is completely shaken up when this happens and the sludge blanket is broken. Until 4–5 p.m. the water leaving the basin may be very turbid. This effect is almost certainly due to the water in the lower part of the basin becoming warmer than the water in the upper part. In at least one case observed this was probably due to the inlet pipe being laid above ground for several hundred metres. The water enters at the bottom of a hopper-bottomed basin and as the morning sun heats the incoming flow the contents become warmer at the bottom, an unstable situation develops and sludge-laden currents rise suddenly to the top.

CHOICE OF SETTLING BASIN

In the discussion of settling basins which follows, the singular when used refers only to the type. No works of any size has less than two basins and most big works have several, the number and size of the units depending in no small measure on the ease with which that particular type can be scaled up.

The object is to choose a basin which will cause the floc to settle out quickly and from which the settled sludge can be removed without too much trouble. Some basins are highly efficient but difficult to clean. Some are ideally suited to small works but do not benefit from scale effect and decline in merit on bigger works. Some work well when handled by experts but are unsuitable for the unskilled. Some basins will cope with higher silt peaks than others. Some are easy to build and some rather complicated. Some take up more space than others.

Although there is a wide choice, there is generally one type of basin that suits a particular job and in Chapter 8 all are mentioned and their merits (or otherwise) considered. Some of the more common types are shown in Table 7.1. This table should be used selectively, as some of the columns may not apply in any given case. In general, the columns on the left will interest engineers in more sophisticated countries and those on the right will be significant in developing areas.

8: Settling basins: types and operation

Following pre-treatment the water enters the basin carrying with it the floc. Floc is fairly fragile and once broken does not re-form properly. It is therefore essential that the incoming channel should be sufficiently generously proportioned to keep velocities down to 30–60 cm/s. If velocities are lower than this, deposition may commence, and if they are higher the floc may be damaged.

HORIZONTAL-FLOW BASINS

Although statistics are not available, there are probably more horizontal-flow basins in the world than any other one type and quite possibly about as many as the rest put together, but this may not be true of all countries. In its traditional form a horizontal-flow basin (Fig. 8.1) resembles a large square or oblong box, filled almost to the top with water. The bottom is flat or has only a slight slope and the water is normally 3–4 m deep. Water enters at one end at or near the top and leaves at the other end over a surface weir. There may be baffles within the main box structure to inhibit short-circuiting. The basins are generally quite big and may have a capacity of 3–8 h of throughput. Smaller ones do exist but are not common.

They are very easy to build and operate. They are not 'temperamental' and will put up with a lot of inexpert handling. Their

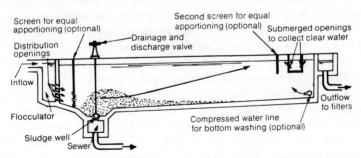

Fig. 8.1. Horizontal-flow sedimentation basin

considerable size makes it unlikely that sudden changes in raw water quality will take their operators by surprise. They 'scale up' very favourably and are at their most economic on big works. They are also at their best when silt loads are exceptionally high: they precipitate silt quite well, have room to store it and are not too difficult to clean. Their cost per unit of volume is low, and although they are bulky in appearance they are normally very cheap in overall cost. They are, therefore, very good performers on big works on silty rivers and can be operated by low grade staff.

Their weaknesses are that they are prone to streaming under certain atmospheric conditions, and they cover a large area of ground. They do not benefit by the sluge accretion effect, and they may need to be preceded by flocculation. In cases where flocculation has been omitted where it would have been advisable, floc can often be seen forming part way along the basin, the size of which still gives it time to precipitate. This is a sign of distress, however, and should direct the management's attention to the merit of adding flocculation as an afterthought.

As they have flat or gently sloping bottoms, horizontal-flow basins are not difficult to clean, but if the raw water is very silt-laden, scrapers should be installed. If the silt is not too voluminous the basins can be emptied down occasionally in rotation and cleaned out by hand, generally with the use of high pressure jets from hosepipes.

On small works, on clean rivers, and where land is expensive, the horizontal-flow basin tends to have been superseded by its competitors and in the UK and Japan it has tended to fall out of favour. It is still very popular in North America because its advantages under typical American conditions outweigh its disadvantages. It is a great favourite in developing and tropical countries, for which it is ideally suited. It is an extremely reliable basin and is very popular with plant operators because it rarely gives trouble and can always be relied on to outperform most other types when the river is carrying a high silt load and good behaviour is at a premium.

Certain features common to earlier basins such as 'over-and-under' baffles and 'ridge-and-furrow' bottoms have fallen out of favour. Both are quite useless. The over-and-under baffle causes sludge pick-up. The ridge-and-furrow bottom has proved incapable of effecting sludge removal. When the perforated draw-off pipes in the bottom of the furrows are operated, narrow waterways are pulled through the sludge and no great volume of sludge is ejected. The ridge-and-furrow bottom has the further disadvantage of precluding the possibility of using a scraper, which is the easiest method of removing sludge.

Design

The principles which govern the design of horizontal-flow basins are discussed in Chapter 6. The design procedure, based on the laboratory approach, is as follows.

Tests should be made to determine the period required for the water to settle naturally in a cylinder equal in depth to the basin, without the addition of coagulants or stirring. In many quite heavily turbid waters the samples quickly show a clear dividing line between the upper clarified zone and the lower zone of settled silt.

If this period is short and the line of demarcation is well defined, flocculation is probably not necessary. If the period is lengthy and the zone of junction is blurred, colloids are probably present and flocculation is essential. A commonly accepted settling velocity for well formed floc is about 3 m/h.

Jar tests should be made on a laboratory scale to ascertain the optimum dosage of coagulants, the flocculation time, and the advisability of using coagulant aids. The cylinder test should be repeated using water to which the coagulants have been added.

The time it takes for the average suspended solids of the water at all draw-off points above the silt line to fall to 2 mg/l should be multiplied by a factor of safety of 3 to arrive at the nominal retention capacity of the settling zone of the basin. The factor of safety allows for inefficiency of the basin due to streaming.

In softening and iron-removal plants, basins of 1–3 h nominal retention are common and effective. For river-derived supplies under typical UK conditions, nominal retention periods are commonly 3–3½ h. On the more silty rivers commonly found in the tropics, where the conditions are otherwise favourable (high temperatures and heavy silt particles) 4 h basins are common, but where conditions are particularly difficult (low temperature, colloids and heavy silt) 6–9 h retention is not uncommon, and in some of the more difficult cases pre-sedimentation tanks and flocculation are also necessary.

If coagulant aids are found to be effective it is probable that the above retention times could be reduced, but these aids need skilled administration and may not be suitable for use in developing countries. Polyelectrolytes are particularly tricky. Incorrect dosing can be counter-productive.

The depth of the settling zone is normally about 3 m; for very turbid water one should add 0.6 m in which precipitated silt can accumulate while awaiting removal.

Although it it no longer fashionable to rate basins in terms of nominal retention time but rather in terms of Q/A (surface overflow rate), it is apparent that if the depth D is fairly constant at about 3 m then the two are directly related. (A depth of 3 m is common for basins up to 60 m in length. Above that a depth/length ratio of 1:20 is commonly noted. In both cases 0.6 m of extra depth may be added as storage space where silt is excessive.)

Table 8.1 shows the overflow velocities that are commonly used.

Factors which favour settlement are coarse-grained sediment, high temperatures and low turbidity; factors which hinder it are colloids, cold water, high turbidity and the coincidence of peak turbidity with peak water demand (cf. Chapter 7). One must look at the worst conditions in each case and decide how bad or easy the situation may be at maximum works output, or alternatively how much water has to be produced when the river turbidity is at its worst.

When in doubt the basins should be made big enough—the extra capacity does not cost a lot and no works ever suffered because of over-sized basins.

Most people believe that long, narrow basins perform better

Table 8.1. Recommended overflow velocities for horizontal-flow basins

Type	Q/A, m³/day per m²		
	Normal conditions	Easy conditions	Very bad conditions
Without coagulant aids	18	24	9
With coagulant aids	27	36	18

than square basins. A length/breadth ratio of 3:1 has sometimes been recommended. There must be some merit in this because length discourages streaming, and also a narrow basin permits the use of certain types of scraper. However, there are a number of quite efficient, absolutely square basins and they are the cheapest to build per unit of capacity.

If sludge is voluminous, mechanical scrapers are necessary. In a long, narrow basin these scrape longitudinally into a hopper at the inlet end. In a square basin the scrapers have a rotary movement and push the sludge into a central hopper. Discharge is under hydraulic head and the fairly liquidized sludge may have a W/V of 2% solids. To avoid settlement in the sludge drains, velocities should exceed 1.4 m/s. If circumstances permit, the sludge should be put back into the river, because most forms of sludge treatment are expensive.

The average velocity in the basin should not exceed 0.02 m/s but in most practical examples the velocity will be well below this.

The water coming into the basin carries floc which is rather fragile and must not be broken up as it does not easily re-form. To ensure that the floc is undamaged the velocity in inlet pipes and channels should not exceed 0.6 m/s and is often slightly less. The inlet channel should run the complete width of the basin to ensure even distribution over the entire cross-section of the basin.

To ensure even flow a perforated baffle wall should stretch across the full width of the basin about 1 m from the inlet end. It should start just above surface and terminate about 1.5 m below. The velocity through any openings should not exceed 0.2 m/s. It should be possible for much of the water to enter the basin by passing down and under the wall.

One of the most important features of any basin is the outlet weir, which is situated at the surface and has a length at least equal to the width of the basin. It has been found that short weirs, with high loadings per unit length, induce currents capable of picking up deposited sludge from the floor of the basin. To combat this, the weir lip length can be increased by making the weir of trough section with water entering over both sides and discharging sideways along the trough. In narrow basins two or more such troughs may be built or the weir length may be increased by indentations. For a weir loading of 150 m³/day per m, which is the maximum favoured, a single-lipped weir cannot discharge the water passing through a basin designed for a surface overflow rate of 18 m³/day per m² if the basin exceeds 8 m in length, which in practice applies to nearly every basin in the business. Multiple weirs of some sort, therefore, are inevitable and although in theory they can be placed almost anywhere, in practice they should be as near the outlet end as possible to avoid fouling the scraper arms and to ensure a fair depth below surface of clarified water.

ALTERNATIVE FORMS OF HORIZONTAL-FLOW BASIN

The rectangular basin may not be the best shape either for scraping or for building cheaply in concrete, and efforts have been made to retain the merits of horizontal flow and eliminate some of the weaknesses. As a result, radial-flow tanks (circular in plan), multi-storey tanks and, more recently, shallow-depth tubed or plated tanks have been designed.

Radial-flow basins

There is no fundamental difference in hydraulic design between the rectangular cross-flow tanks and circular-shaped radial-flow tanks.

In a radial-flow basin the raw water enters through a central inlet and flows radially outwards towards a continuous peripheral outlet weir. The same values for surface overflow rates are applied and much the same results are obtained. Obviously, radial flow velocities cannot be uniform because the cross-section increases with the radius, but this is not necessarily in itself a weakness, as maximum cross-section and therefore minimum velocity occurs where it is most needed, which is after the more rapidly settling particles have deposited.

Small circular tanks tend to be cheaper to build in concrete than square tanks of the same capacity and lend themselves to pre-stressing, but they cover more ground per unit of area because of the waste space caused by the circular shapes. The outlet weirs present less of a problem because they can be placed right round the outer edge of the tank. Scraping is easier because scrapers with a rotary movement are relatively cheap, and effectively clear all the sludge.

However, streaming is a particular nuisance in circular basins because they are basically a bad hydraulic shape, length (i.e., radius) being less than breadth (i.e., perimeter), and circular baffles normally have to be installed. For this and other reasons they have never been as popular as rectangular basins.

Inward-flow tanks (with the water entering at the perimeter and

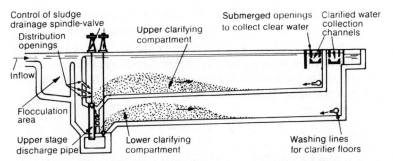

Fig. 8.2. Multi-storey basin (the storeys working in parallel)

flowing to a central outlet) are virtually never used because many of the merits described as applying to the outward-flow basin are lost or even reversed. Ring-shaped basins in which the water completes a circular track from the inlet to outlet weirs placed back to back along a radial division wall are not favoured owing to the varying length of travel of the water. They are very rarely encountered.

Multi-storey tanks

Where space is limited, or for structural cheapness, there is no reason why a conventional horizontal-flow tank cannot be built in the form of a structure of two or more storeys (Fig. 8.2). There are several known cases where the output from a basin has been doubled by the construction of a floor at mid-depth in a conventional tank, which (by halving the effective depth) doubles the effective area. As area is such an important factor in settling, multi-storey tanks are often remarkably cheap and effective. Scraping can sometimes be confined to the upper chamber if all the water passes through the storeys in series, because most of the deposits form at the inlet end of the top storey. These basins are fairly common in the eastern states of the USA and in SE Asia and generally perform impressively.

A recent example built by the French firm of CTE at Pakse (Laos) was designed to treat 7500 m³/day with the following criteria:

flocculators —detention time 20 min
settling tanks—detention time 105 min
——total capacity 585 m³
——useful capacity 520 m³
——number 2
——storeys 2

This works performed well on Mekong River water which is not particularly difficult to treat (although at several installations on the river, trouble occurred due to abrasion of the raw water pump impellers). Flow through the two-storey basins was in parallel, and in such a case one basin might have to be shut down for cleaning and the flow deflected through the other.

Plated tanks

Hazen and Camp arrived at the logical conclusion (cf. Chapter 6) that if area was the main consideration in settling tank design, the more extensive and shallower the tank the better. The practical limitations of this theory would be imposed only when the velocity through the tank reached a value at which the deposited silt could be lifted. However, few (if any) very shallow tanks of great surface area have been built, but in recent years this principle has been adapted to the design of plated tanks and tube settlers.

The plated tank is shown in Fig. 8.3. The plates are supported by hangers and dog-legged in the manner indicated. They are in sections rather less than 1 m² in area and each makes an angle with the vertical of 60°. In operation the silt deposits on the plates and slides down, from plate to plate, through the gaps, on to a belt scraper and can thus be ejected from the basin.

Practical design considerations are that one third of the tank at the

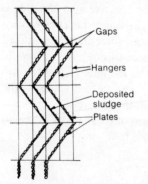

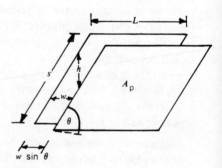

Fig. 8.3. Japanese plate settler: vertical section (*the direction of flow of water is at right angles to the plane of the page*)

Fig. 8.4

inlet end should act as a conventional horizontal-flow basin, and remove the coarser particles; that flocculation by paddle is essential and coagulant aids should be widely used; that the plates should be placed so that the tops just break surface and the base of each bank is about 1 m above the floor; that a low cross-wall on the basin floor should deflect all flow through the plated section; that velocity through the plated area should not exceed 0.8 m/min; that the time of passage through the plated area should be about 20–30 min; that the plates should be 8 cm apart horizontally; and that the scraper should be of the chain-link belt type.

These basins can be very small, but in an observed example at Seoul, Korea, the cost of the basin plus all its necessary flocculators, plates and scrapers was comparable with that of a much bigger conventional basin working on the horizontal-flow principle. The Han River above Seoul (cf. Appendix 1) is normally fairly clean, but at the time of the spring break-up shows high turbidity which has to be treated under near-freezing conditions.

This plant is of Japanese origin and is similar to a basin designed for use at the Shiomidai plant at Kawasaki, Japan.[1] This type of settler is widely used in Japan, but Japanese rivers are normally fairly clear and there is little experience as yet of its performance under adverse silt conditions. The design of the plated tank is covered by Japanese patents.

Detail design. As this type of basin is rather unusual, some notes on the design of the plated section are given (by courtesy of the Waseda Co. Ltd, Tokyo).

At Seoul the rate of flow Q through the tank was 250 000 m³/day (10 416 m³/h) and the assumed settling velocity of a well-flocculated particle was 0.6 m/h. For $Q/A = 0.6$ m/h (a conservative value for alum floc), the effective area A of plate required is 10 416/0.6 m² = 17 361 m².

Referring to Fig. 8.4, the quantity treated is

$$Q = A_f v_f = (Sw \sin \theta)(v_s L/h)$$
$$= (SL v_s \sin \theta)/\tan \theta = A_p \cos \theta \, v_s$$

where v_f = flow velocity
$\quad A_f$ = flow area
$\quad S$ = slant height of plate
$\quad w$ = horizontal distance between plates

h = vertical distance between plates (= settling distance)
v_s = settling velocity
L = length of plate
A_p = plate area.

The plates at Seoul were 1 m by 0.92 m, area 0.92 m². The effective settling area for each plate was 0.92 cos 60° m² = 0.46 m². If N is the required number of plates, and 0.7 is the 'efficiency' of the plates,

$$N = \frac{A}{0.46 \times 0.7} = \frac{17\ 361}{0.46 \times 0.7} = 53\ 916.$$

Where the sloping plates are arranged in four vertical tiers and in ranks as shown in Fig. 8.4, the number of sloping plates required in one row and one rank (one layer) is $N/4$ = 53 916/4 = 13 479. When the pitch of the sloping plates is 80 mm (Fig. 8.5) and the basin is 70 m wide, the number of plates in one row is 70/0.08 = 875. The number of ranks is 13 479/875 = 16.

The actual total number of sloping plates is 875 × 4 × 16 = 56 000; settling area = 56 000 × 0.46 × 0.7 m² = 18 032 m².

The area through which the water flows is 70 m × 3.348 m = 234.36 m². The mean velocity of flow v_f = 10 416/234.36 m/h = 44.45 m/h = 0.74 m/min. The settling distance (Fig. 8.5) is 80 tan 60° mm = 80 × 3$^{1/2}$ mm = 138.56 mm. When the settling velocity is 0.6 m/h, horizontal arriving distance is x = 44.45 × 0.13856/0.6 m = 10.264 m.

It has generally been observed that the calculated distance should not exceed 75% of the actual distance, and at Seoul the 16 ranks provide 16 m of distance, giving a factor of 10 264/16% = 64.15%, which would be considered satisfactory.

Settling time T is total length of plate divided by velocity of flow, giving 16 m/44.45 m/h = 0.36 h = 22 min. This was the nominal

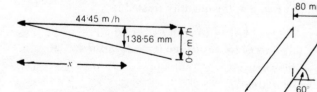

Fig. 8.5

retention time at Seoul in the plated section, to which should be added the times spent in the flocculation tank, the one third of tank area provided before the plated section to act as a preliminary basin, and an additional section to accommodate the belt type scraper.

Suitability. The plate settler would appear to have as much potential as the tube settler in upgrading the throughput of existing conventional basins. In the above example, the design assumes permissible horizontal flow velocities of 0.74 m/min, and the settling area is much greater than in a conventional tank. Therefore the output capacity of a conventional tank could be greatly improved by the insertion of plates.

The plated clarifier is a sophisticated piece of equipment and the same description can be applied to several of the high efficiency basins described below. All are capable of impressive performance but all depend on first-class operation and the question must invariably arise as to their suitability in the particular cirumstances for which they are being considered. In many instances the conclusion must be reached that their adoption might be delayed with profit, and a simpler basin installed for the time being.

Tube clarifiers

At first glance it would appear difficult to know whether tube clarifiers should be classified as upward- or horizontal-flow basins because the flow appears to be moving more vertically than horizontally. However, the action is essentially that encountered in a very shallow horizontal-flow plant.

Fundamentally, the action takes place in 5 cm square tubes inclined at 60° to the horizontal, the water passing upwards through the tubes (Fig. 8.6). Clarification depends on the settling capacity of a particle in what is effectively a horizontal-flow basin only 5–10 cm in depth. While settlement is satisfactory and would best take place in a horizontal tube, the problem of silt removal becomes paramount. For this reason it is essential that the tube should be set on a slope. A vertical tube would have no advantage whatsoever because it would behave in the same way as an ordinary upward-flow tank, but it has been found[2] that if the slope is about 60° to the horizontal, horizontal-flow principles apply and the sludge trickles downwards and keeps the tubes clean.

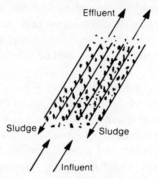

Fig. 8.6. Tube settler

In practice the tubes are racked up in shop-fabricated panels and banks of these panels are placed together so that water enters at the base of the tank and passes upwards, depositing the silt inside the tubes, with the clarified water emerging at the top of the tubes and being collected at the surface by a system of launders. The tops of the tubes are placed 60 cm below the surface of the water in the tank. Loading (Q/A) is normally 9 m^3/h per m^2 (216 m^3/day per m^2).

Perfect flocculation and the use of coagulant aids must be assumed. However, in terms of the normal velocities encountered in an upward-flow tank, something approaching three times the loadings can be attained, and compared with horizontal-flow tanks of comparable output the permissible surface loading may be 4–5 times as great. Against this must be set the additional cost of the link-belt type scrapers required plus the cost of the tube racks. Studies indicate that the tube clarifier shows a small overall cost advantage.

It is perhaps a reflection on the innate conservatism of the water engineer that only a few of these tanks have been built, in comparatively small sizes. Many of these are in the southern states of the USA. It has been suggested that this type of installation could be brought to a halt by freezing, but as a number are working in Canada this cannot be unduly significant.

Perhaps more headway must be made elsewhere before tube clarifiers can be considered in developing countries.

The most effective use of this type of plant might result from the insertion racks of tubes into existing basins in order to uprate the capacity where funds or space preclude a completely new installation. Most types of conventional basin lend themselves to this sort of conversion. Taking the case of a normal (4 h) horizontal-flow basin it would be possible to upgrade its output approximately 4–5 times by installing tubes in the middle half of its length (using the first section of its length as a preliminary grit chamber). Such applications are now gaining in popularity and would appear to have a considerable future. In all such applications, however, it should be remembered that a basin generally fails when sludge is depositing faster than it can be ejected. The sludge ejection capability should always be checked before a scheme is adopted to increase deposition.

The Lamella separator

The Lamella separator[3] (Fig. 8.7) is a type of plate settler particularly common in Holland, Sweden and West Germany. It operates more or less on the same principle as the inclined tube but the turbid water enters at the top and flows downwards instead of vice versa. This downward component of flow assists in the ejection of the sludge and permits the plates to be inclined at a smaller angle (30°) with an interplate clearance of 35 mm.

For every 1 m² of top open area, a projected area of 10 m² is

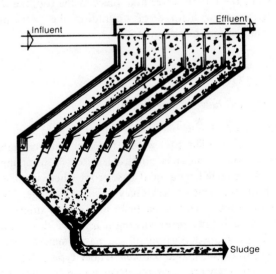

Fig. 8.7. Lamella separator (by courtesy of the Parkson Corporation)

available, so the capacity may theoretically be ten times as great as that of a conventional horizontal-flow basin of similar size, but the qualifications about pre-treatment and overall cost apply as above.

As the clear water is travelling downwards, in the same direction as the sludge, some special method has to be used to separate the one from the other at the bottom of the tube. In Fig. 8.7 the special U shaped collectors and clear water return tubes can be seen. In fact the clarified water exits upwards and from the top of the separator.

In a quoted case from West Germany,[3] the incoming water had filterable matter of 4.5–478 mg/l and pH 7.0–8.3. The chemicals added were ferric chloride 75–100 mg/l, lime 27–35 mg/l and a floc aid 0.5 mg/l. Flocculation time was 30–36 min. Based on top area of the tubes, Q/A would be 490 m/day. The effluent had suspended solids of 2–9 mg/l.

The water being treated may have been particularly favourable but even so the published results are impressive. The lower angle of inclination compared with the normal tube settler exposes a greater area for settling to take place per unit length of tube and theoretically the Lamella design should be more efficient. The case quoted would appear to confirm this. However, as both the entry of the raw water and the exit of the settled water take place at the top, the use of the Lamella settler as a method of upgrading the performance of existing tanks appears to be limited.

UPWARD-FLOW BASINS

Hopper-bottomed sludge blanket basins

Hopper-bottomed sludge blanket basins (Fig. 8.8) are deservedly popular in the UK. The raw water (together with coagulants and frequently coagulant aids) is admitted, after flash-mixing, at the bottom of the inverted pyramidal base from where it passes slowly upwards through a zone of previously deposited sludge. This acts to flocculate and entrap the floc particles and greatly improves clarification. When such tanks are working a layer of clear water in the upper cubically shaped portion of the basin makes it possible to observe the top of the sludge blanket. Under typical UK conditions they work well. They are normally designed on the basis of the rate of rise permissible in the upper cubical zone where the area is at a

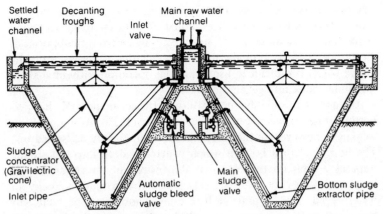

Fig. 8.8. Hopper type settling tanks

maximum in plan. They were originally designed for, and are particularly effective on, water softening plants. For such applications the permissible maximum upflow rate may be as much as 4–4.5 m/h.

Where coagulant aids are used even river waters can be treated successfully at upward velocities of about 3 m/h, but where these are not used velocities of 1.5 m/h are rarely exceeded. All the velocities quoted above contain a factor of safety of 2, and depend on the theories described in Chapter 6, wherein particles are assumed to fall at least twice as fast as the column of water is designed to be rising.

The great merit of these basins is that, under conditions that suit them, they can be regarded as providing mixing, flocculation and settlement and they give a settled water of great clarity. Deposited silt can be removed easily without a scraper, using only the available hydraulic head. In fact the ease with which sludge can be ejected is perhaps their best feature.

In the side of each basin at a sill depth of about 1.2 m below surface it used to be common practice to form a 'pocket' in concrete into which the silt could decant on reaching the sill height and from where it could be drawn off as a highly turbid liquid through a small (20 mm) pipe into the drain. Fig. 8.8 shows a more sophisticated type of sludge draw-off, known as a Gravilectric cone. Such cones are described below, under 'Modified hopper-bottom designs'. They perform the same function as the old concrete pockets but lend themselves to automatic control and operation. In several cases where they have been installed in existing tanks, substantially improved performances have been noted. Should complete emptying of the tank be required a 100 mm pipe drawing from the bottom of the hopper can be brought into play, but this is rather a drastic device which is difficult to use in normal operation without shaking the sludge back into suspension.

Because of the highly successful experience in the UK, hopper-bottomed basins have been widely used abroad. They contain a basic weakness, however, in that they cannot economically be 'scaled up'. The hopper shaped bottom which penetrates deep into the ground imposes a limit on the economic dimensions of the square top (a limit which arrives more quickly when the ground is rocky or waterlogged), and in practice bigger installations can be built only by adding an increasing number of fairly small units of standard size, each not exceeding 10 m on the side. While a reassuringly small total area of basin can still appear on the design drawings, it consists in most cases of a great amount of awkwardly shaped concrete, and the cost is commensurately high. It therefore follows that the hopper-bottomed, upward-flow tanks are most suitable for smaller works, of perhaps less than 45 000 m³/day.

If the job is located overseas, other difficulties arise. This type of basin is not common in North America, and in fact has tended to proliferate where the British have journeyed. The basins are not completely at ease when encountering the high silt loads found so often in tropical rivers, purely because they are small and tend quickly to become silt-logged unless the small-diameter pipe bleeds are considerably increased in size. At the Johore works in Singapore, where 2 in pipes had to be installed, water losses from the basins were apt to rise as high as 15% of the thoroughput when the works was initially on test, although this was subsequently improved.

At the Sarrafiyah works in Bagdad this type of basin was frequently in trouble during periods of high turbidity, until a preliminary basin of 1 h capacity (not using coagulants) was built in line ahead of it to take the occasional shock silt loads the Tigris could produce. This preliminary basin owed little to theory: it was simply the biggest 'box', square in plan, that could be built on the limited space available. It worked extremely well and at the time the general opinion was that it could have been made smaller. In 1977, however, very turbid water overwhelmed the upward-flow (Pulsator) basins at Tehran, even though these were protected by radial-flow preliminary basins of 1 h nominal capacity. The suspended solids in the river were quite exceptional (57 000 mg/l) and the radial-flow basins were probably short-circuiting to some extent: the treatment plant was put out of commission for about 12 h. As a result of this happening, local informed opinion was that the preliminary basins should have been anything up to 2 h capacity to have withstood the extreme conditions. In the event, however, although the water level in the city's service reservoirs fell dramatically, little effect was felt in Tehran itself.

There is one final cause for concern regarding the use of hopper-bottomed upward-flow basins, and that is the suspicion that in hot countries the water they contain is, in certain circumstances, liable to 'roll over' every day early in the afternoon (cf. Chapter 7). This tendency has been reduced by the introduction of Gravilectric cones which have improved control of the sludge blanket.

Modified hopper-bottom designs. The earlier forms of hopper-bottomed basins, as described above, had inherent weaknesses. They were expensive to build and difficult to operate on silty rivers, and they could have high water losses. However, they operated on a principle which was basically desirable—the upward passage of water through a zone of mature floc particles which captured and retained the fine particles in the incoming water. Recent developments have resulted in simplification of the basins while retaining the sludge blanket principle.

Three changes have been introduced: the multi-hopper tank, the flat-bottomed tank with vertical sides, and the Gravilectric cone. The multi-hopper tank is operationally a single tank with linked hoppers: it is equivalent to a battery of hopper-bottomed tanks with the upper parts of the walls omitted.

The flat-bottomed tank with vertical sides is similar to the type built at Singapore and also at Newcastle upon Tyne some years ago, which can be designed to be fitted with floor silt-scrapers. Several have been built (e.g., at Lisbon, Ankara and Tauranga (New Zealand)) for treating river water at upward flow rates of 3–3.5 m/h using alum and polyelectrolyte. It would appear, however, that most of these examples are dealing with water of less than 500 mg/l suspended solids.

The Gravilectric cone is a PVC-impregnated nylon cone suspended within the clarifier with its rim below surface to suit the level of the floc blanket. Excess sludge flows into the cone, whose weight in water is so low as to permit the weight of sludge to be accurately determined. The area of the top of the inverted cone is large and control of the sludge blanket is ensured. When a pre-determined weight of sludge has filled the cone a switch opens the sludge bleeds, concentrated sludge is released and loss of water is minimized. Where Gravilectric cones have been fitted on existing works in the UK the average rates of upflow have been doubled from 1.2–1.8 m/h to 2.4–4.2 m/h, and the water losses reduced from 2.5–7.0% to 0.1–2.0%. The Gravilectric cone has been developed and patented by Paterson Candy International.

Accelator type solids contact clarifiers

The Accelator type of clarifier (Fig. 8.9) is made by many manu-facturers under different trade names (Accentrifloc, Clariflocculator etc.) but the original patents were held by Infilco Inc., USA, who are also suppliers of the plant. This type of plant uses the principle of sludge accretion while retaining the great advantage of the flat bottom. It is circular in plan and may be built up to a little over 30 m in diameter. The conical section in which the sludge collects is formed by concrete walls within the tank and this eliminates the necessity to dig deeper to accommodate the inverted pyramid when bigger tanks are needed. This feature therefore permits this tank to be built in bigger sizes without sacrificing economy. There are those who believe that this type of basin represents the ultimate in cheapness and effectiveness on big works.

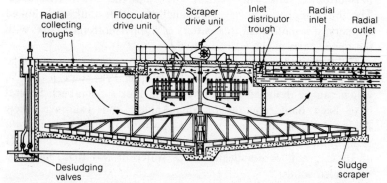

Fig. 8.9. Accentrifloc clarifier (by courtesy of Paterson Candy Inter-national Ltd)

In shape it is circular with a flat bottom. Internally, it has a conical hood which divides the tank into two zones. In the inner zone, raw water and coagulants are mixed and are driven by a centrally mounted impeller into the outer chamber, which acts much in the same way as the upward-flow hopper-bottomed tank described above. The 'hopper' is formed by the shaped hood which forms the inner wall of the annular space. The surface area of this space is dimensioned to provide the same upward flow velocities which control the dimensions of any upward-flow tank. Water is normally drawn off over radial launders spaced at regular intervals to ensure uniform upward velocity.

The Accelator depends on the recirculation of pre-formed sludge which can be readmitted through trapdoors from the bottom of the hopper to the central primary reaction zone, or, if surplus to requirements, ejected from the central section to the drains. This type of plant can be designed for rise rates up to 5 m/h, but without coagulant aids it commonly operates at about 2.5 m/h. It requires more skill to operate than the basins previously mentioned as it depends to a marked degree on the speed of the impeller and the correct sludge concentration. The impeller speed is normally higher for heavily turbid waters than for water that is merely coloured, and the speed is generally determined by visual inspection of the peripheral clarification area. If clouds of floc particles appear the speed must be reduced. The same applies to the central zone where undue turbulence denotes that the turbine speed is too high, and patches of clear water denote that it is too low.

As for all sludge blanket tanks, the action takes time to establish; when starting up, the trapdoors at the bottom of the hoppers must be left open until enough sludge has formed for the process to work. When the tank is working normally, sludge ejection must be maintained at a rate which balances sludge production and does not lower the sludge concentration to a point where the effectiveness of the process might be impaired. The top of the sludge blanket can normally be observed through the clear water of the peripheral zone and should be seen to be in the lower part of the annular zone. If it tends to rise above this, the sludge concentration in the central zone is probably too high (above 10%) and it must be regulated by increasing the rate of sludge ejection. If the sludge concentration is too low (below 4%) either the impeller speed or the coagulant dose is too low.

The sludge concentration is measured in a graduated cylinder and is equal to the amount of wet silt by volume after 10 min settlement.

Possibly because of the constant recirculation of water these tanks do not 'roll over' under tropical conditions. Under constantly changing river conditions they need close observation to ensure that impeller speed, trapdoor opening and sludge concentration are correctly set to maintain optimum performance. Like all high efficiency basins they do not cope particularly well with water of excessive turbidity, and require quite expert handling. The concrete structure is complicated to build and consequently rather expensive. Where the raw water is variable in quality and the attendants of less than average capability they have not proved universally successful. However, given reasonably skilful control they have done well and are performing satisfactorily in Iraq, Tehran, Kuala Lumpur, Singapore, Hong Kong and other places. They are now tending to be superseded by other types of basin but there are many existing installations.

The Pulsator

The Pulsator (patented and made by Degrémont) is another type of upward-flow tank which depends on a sludge blanket for its effectiveness. It also combines the merit of having a flat bottom with the operating simplicity of the hopper-bottomed tank. It can thus be scaled up into large sizes economically.

The Pulsator differs from the normal sludge blanket tank in that water is admitted at varying rates of inflow, a distinct surge being succeeded by a period of quiescence (Fig. 8.10). Rather like the sand in a filter during the washing cycle, the sludge blanket expands during the period of maximum inflow and contracts as soon as inflow diminishes. The design is such that the speed of inflow is not allowed to exceed limits which would break up the blanket. The gentle up-and-down movement induced in the sludge blanket creates a thicker, more uniform sludge zone which improves the clarifying action. The sludge is decanted over a weir placed at about half tank height, and as the receiving tank has a hopper bottom and there is no water movement therein, the sludge tends to concentrate and can be ejected easily under normal conditions. If the sludge is very heavy, however, ejection can be a problem.

As in all upward-flow tanks, the water is collected by launders

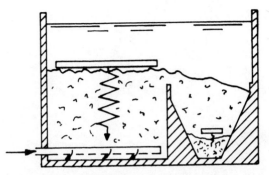

Fig. 8.10. Pulsator: principle of operation; the 'piston–spring' effect is obtained by the pulsing movement induced by variation in flow (no actual 'piston' is installed, of course)

placed at regular intervals across the top to eliminate inequalities in the upward rate of flow of the water. In practice these are generally about 1 m apart and have V notched edges to permit more accurate vertical placement.

The tanks are generally rather deep, 5 m being a dimension commonly used. Their square shape and flat bottom, however, lend themselves to cheap construction, and the effective use of the bucket type of sludge ejector makes for simple operation.

The improvement in the sludge blanket caused by the gentle pulsing effect permits upward water velocities of up to 6 m/h, which makes the Pulsator one of the most efficient of basins. This velocity figure should be regarded as a maximum. There are many examples where it is lower. At the Kan works in Tehran the upward velocities in the Pulsators average 2.4 m/h, but no coagulant aids are used.

The variation in flow rate is achieved by flow into and out of a vacuum chamber. The additional cost of providing and operating the vacuum pumps and chambers has to be taken into account when assessing the overall cost.

There are many examples of the successful use of this sort of basin and in underdeveloped countries its simplicity and effectiveness have won it many supporters. However, it is not a very easy basin to clean because the bottom is obstructed by pipes. The continuous sludge ejection deals with a high percentage of the sludge, but some accumulates near the inlet pipes, making it necessary to shut the basin down occasionally to permit manual cleaning. Where the sludge is volu-

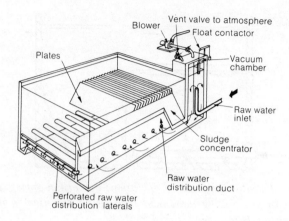

Fig. 8.11. Super Pulsator (by courtesy of Degrémont)

minous this is essentially a basin which benefits from the installation of pre-settling. For the same reason it is not frequently found on softening or iron-removal plants.

A Super Pulsator (Fig. 8.11) is available which has plates installed in the upper part of the basin to introduce a shallow-depth settling action actually within the sludge blanket.

Singapore-type upward-flow tanks

A type of tank built at the Woodleigh works at Singapore has done extremely well. It is a flat-bottomed upward-flow tank designed to work with flow rates comparable with those used in other tanks working on this principle. Water is introduced at the bottom through horizontal perforated pipes, over which continuous inverted V shaped cowls have been placed to distribute flow more uniformly. The clarified water is withdrawn over surface launders in the usual way. It is not unlike a Pulsator without the pulsing movement, but (again like the Pulsator) cleaning by scraper of sludge accumulating at the bottom is impossible and the basins are shut down in rotation and cleaned manually.

These basins were designed and built in 1954 but other considerations intervened and filters were added only in 1969. The water treated is generally fairly clear and mostly from storage reservoirs on the island. The upward velocities as designed were 4 ft/h (1.2 m/h) and the coagulants used were aluminium sulphate and lime.

A horizontal-flow basin originally built at the Ayer Itam works at Penang was converted to this principle by the then water engineer Mr Goh Heng Chong to treat reservoir-derived water and a marked increase in capacity was achieved.

Pretreator

The Pretreator (Fig. 8.12) combines mixing, flocculation, clarification and high rate sludge recirculation in a single tank which also has the structural merit of a flat bottom. Raw water previously mixed with concentrated sludge passes up the centre column and enters the flocculation zone. Settlement takes place in the clarification zone. Solids are transferred through a pipe back into the sludge recirculating well to mix with the incoming raw water to effect continuation of the process. Surplus sludge is drawn off through a pipe in the bottom of the tank floor, and the clarified effluent leaves over a peripheral weir at surface level. Mixing and flocculation are accomplished without mechanical equipment. No sludge blanket is formed. Rapid sludge recirculation is the key to the high efficiency of these basins.

The sizes available range in diameter from 30 ft (9.17 m) to 100 ft (30.5 m) and the capacities from about 3000 m³/day to 45 000 m³/day. Upward flow rates of about 2 m/h are commonly found in the clarification zone. The Pretreator is recommended for water with up to 1000 mg/l total solids.

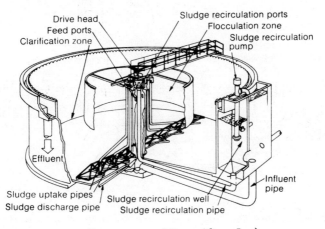

Fig. 8.12. Pretreator (by courtesy of Dorr Oliver Inc.)

Many of these tanks have been built in the USA but they are simple to operate and they are now being used in India and Zambia. In Egypt the Pretreator is being successfully promoted to treat the Nile water, the characteristics of which have been radically altered by the construction of the High Dam at Aswan.

Circular upward-flow basins without sludge recirculation or marked sludge blanket effect

The Centrifloc type of upward-flow basin (Fig. 8.13), without sludge recirculation or marked sludge blanket effect, is of simpler shape than the Accelator, merely consisting of a circular tank inside which a smaller circular inner compartment acts as a flocculation chamber. The flocculated water passes downwards through the central zone and escapes radially outwards underneath the dividing wall (which is suspended from above and nowhere touches the floor). The outer, annular, compartment is sized to keep upward velocities at the levels acceptable for upward-flow tanks and the water is decanted from the top by radial collecting troughs. The floor is generally constructed with a slight slope towards a central drain. A centrally driven rotary scraper, with blades set at angles, pushes the sludge towards the drain, from which it can be drawn as required. The scraper sweeps the entire floor area, its horizontal arms being free to pass under the suspended dividing wall. The capacity of this basin to handle heavy silt loads is one of its main features.

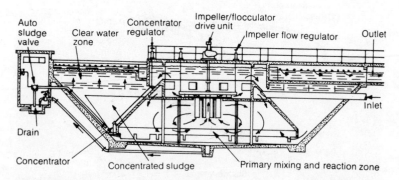

Fig. 8.13. Centrifloc clarifier (by courtesy of Paterson Candy International Ltd)

SPIRAL-FLOW BASINS

'Alexandria pattern'

A circular spiral-flow basin was designed to treat a very difficult water taken from the Sweetwater Canal in Egypt, which is fed by the Nile. Nile· water is at its most heavily turbid and difficult to treat in August (coinciding with peak demand in the city of Alexandria). The passage of the Sweetwater Canal through the heavily populated delta results in the heavier solids depositing and the remaining colloids becoming coated with sewage and extremely resistant to settlement. Normal horizontal-flow basins of as much as 8.9 h nominal retention capacity at Alexandria's Rond Point works had seasonal problems, being prone to streaming caused by wind and temperature. After much experimental work a circular tank was designed for installation at the Siouf works in which well flocculated water was admitted tangentially at the bottom, followed a rising spiral path and was drawn off over a peripheral weir at the surface, extending for about one third of the total circumference[4] (Fig. 8.14). Streaming was virtually eliminated as the incoming water circulated at low level in the basin.

Very favourable results were claimed for these tanks. There was no sludge accretion effect but the settling action clearly owed something to both the horizontal and the vertical components of the spiral path followed. The nominal retention time was said to be 4.25 h.

These basins have had considerable publicity and a few have been built elsewhere, but a close study of the figures quoted[4] fails fully to

Fig. 8.14. Alexandria spiral-flow tank

prove the results claimed. Many questions regarding comparative conditions were raised[5] and these were never answered. This was quite understandable as Alexandria was at that time severely affected by the North Africa campaign of World War II, but the case has always remained a little suspect and the figures are thought by some to indicate that the performance of the horizontal-flow basins of traditional pattern installed at Rond Point was certainly no worse.

On the figures given a 93.3% reduction in suspended solids was attained by the spiral-flow basins at Siouf works when they were being worked at a rate equivalent to 7.42 h nominal retention. On the same water at Rond Point the horizontal-flow basins were producing the same results when working at a slightly faster rate of 6.75 h retention. The better results claimed for the spiral-flow basins appear to have been attained at other times on less turbid water.

A typical horizontal-flow tank rarely works at better than 33% of its theoretical efficiency and although it was claimed that the Alexandria spiral-flow tanks were nearly twice as efficient, some doubt has persisted. Certainly, comparatively few further examples of these tanks have been built. However, even if the results quoted were a little optimistic, the principle on which these basins were built is clearly quite sound and the basins undoubtedly work effectively.

Spiractor

The use of the Spiractor (made by Permutit-Boby Ltd) is confined to the lime softening process. It normally takes the form of a steel pressure vessel (Fig. 8.15) shaped like a steeply sided inverted cone, but could equally well have an open top. Raw water is injected tangentially at the base and mixed with the lime solution. At the same time 'catalyst granules' (sand) are injected with the raw water. The reaction takes place in a state of swirling agitation and the precipitated calcium carbonate (from the normal reaction $Ca(HCO_3)_2 + Ca(OH)_2 \rightarrow 2CaCO_3 + 2H_2O$) crystallizes at once on to the granules. The Spiractor is characterized by its small size (about 10 min retention capacity), by the comparative absence of fines in the carry-over, and particularly by the hard and easily disposable granular residue. It is the high settling velocity of these heavy pellets which permits the very high upflow velocities and thus the compact dimensions of the Spiractor.

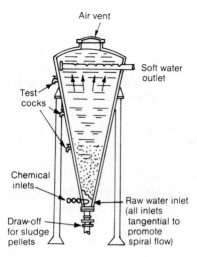

Fig. 8.15. Spiractor

The total amount of sand admitted is regulated, as is the size of the 'grown' grains. The volume of new sand put in daily is only a fraction of that of the pellets ejected. Control is simple and, where conditions suit them, Spiractors give no trouble. In the English eastern counties there are many working successfully, softening water from the chalk. At Swaffham a plant processing 465 m³/day was using 76 kg of local sand and 204 kg of lime per week. One part of tri-sodium phosphate (TSP) was added with each 700 parts of lime, and Calgon was added between the Spiractor and the filters.

Spiractors work excellently when the hardness is predominantly caused by calcium carbonate but are not to be recommended where magnesium is present to any extent (>25 mg/l) because magnesium does not react as quickly as calcium to form pellets.

The lime is added as an emulsion and inevitably calcium carbonate fines are produced, which should be removed by filters, generally of the pressure type. If permanent hardness is present the Spiractor can be followed by a base-exchange softener, and there are successful examples of this type of installation.

FLOTATION

The principle on which flotation tanks work is explained in Chapter 6.

One of the more successful processes relies on recirculating about

7% of the treated water supersaturated with air at 5–6 atm. It is claimed that because air bubbles rise faster than floc sinks, the retention time within the basin may be radically reduced. However, the cost of the air compressing and entraining devices is appreciable and this has to be set against the reduced capital cost of the basins, as have the maintenance and operation costs of the machinery against the chemical savings claimed. The power consumed is said to be about 0.4 kW h per cubic metre of water processed.

It is clear that these basins must have particular merit when separating out algae which tend to float anyway.

Flotation can be introduced into existing sedimentation installations and remarkable results are claimed for some of these conversions. At a plant in Sweden, the output was increased three times, the time between filter washes quadrupled and the alum dose slightly reduced.

BASIN FAILURES

When the silt content in the river is excessively high, upward-flow basins are somewhat more prone to failure than those operating on the horizontal-flow principle. A basin may fail for either of two reasons:

(a) that it is failing to precipitate the suspended solids; or
(b) that it is failing to eject the solids precipitated fast enough and is becoming silt-logged.

The former problem may affect any type of basin but it may be more prevalent among horizontal-flow basins: it is generally due to incorrect coagulant dosing, poor flocculation or streaming. It can be a nuisance but is rarely catastrophic. The latter problem is much more likely in the smaller, high-efficiency, upward-flow basins and it can cause a complete works shutdown. Once an upward-flow basin gets full of silt it becomes unmanageable until the river settles down and the basins and filters have been washed clean. As washwater tanks are limited in size, being designed for normal operation, there is often, in emergency, a loss of washwater and a disastrous chain reaction is apt to run through the works. Operating trouble on a waterworks nearly always originates in the basins, really serious trouble in upward-flow basins. The remedy is provided by pre-settlement tanks (Chapter 3).

SUITABILITY OF PLANT

There is no doubt that the figures claimed for the many upward-flow tanks of proprietary design can be achieved given good conditions, expert control and the use of polyelectrolyte. It is in this direction that settling basin design in developing countries must progress. At the same time it must be appreciated that even in waterworks like that at Tehran, operated by a highly skilled staff with excellent laboratory facilities, lower rise rates than the maximum claimed are being attained because of adverse river conditions.

Simple but well designed works operated by competent engineers (but with few waterworks chemists) are working well with the rather lower rise rates mentioned in Tables 6.4 and 8.1, and it could be argued that this represents the best solution at present for developing countries.

There are unfortunate examples of highly sophisticated works in big towns in Asia, and even capital cities, where an absence of chemists and even chemicals is producing zero performance. Under such conditions no blame can attach to the resident staff, whatever their individual capacity. It is clearly important that waterworks designers should be aware of the potential operating deficiencies before providing plant which might never function.

Those circumstances apart, abnormal river conditions will still remain as the bane of a water engineer's existence. Upward-flow tanks decline in performance when solids in suspension reach freak seasonal highs and they have to be protected by horizontal-flow pre-sedimentation basins. When the first flood of the season carries off the accumulated dust of a tropical summer, or a jungle catchment is cleared for development, or a gelam soil in Malaysia turns the local river water into dilute acid, the local water engineer with horizontal main basins behind his horizontal preliminaries will sleep sounder than most of his less fortunate colleagues.

9: Filtration

Filtration follows settlement, and to a certain extent they are interchangeable; the more effective the settlement the less the filters have to do and vice versa. It is the final process in water clarification and, except in the case of clear underground sources where it is unnecessary, there is a tendency nowadays to regard it as essential. However, there are some major cities, even in highly developed countries, where filtration of surface sources is not practised and the end product is acceptable. In this connection it should be remembered that the WHO international standard for Highest Desirable impurity levels requires a turbidity not exceeding 5 JTUs and that for Maximum Permissible impurity levels requires a turbidity of less than 25 JTUs. There are many surface waters which, particularly after storage, would meet these standards, but they are not very high standards. As time goes on fewer communities of any size will be prepared to accept anything other than crystal clearness in their drinking water supplies, and filtration, certainly of surface-derived supplies, will become virtually universal.

Basically the process of filtration consists of passing the water through a bed of sand, or other suitable medium, at low speed. The sand retains suspended matter while permitting the water to pass, and the filtrate should be clear and sparkling in appearance. There are practical limits to the capacity of filters to achieve this final degree of clarity unless the incoming water from the settling basins is itself of fairly low turbidity, the degree of which may vary with the type of filter adopted (cf. Chapter 2 and below).

TYPES OF FILTER

The filters most commonly found are the rapid gravity sand type. They are normally operated with coagulants and very often follow settling basins. They ought to produce excellent results with incoming water of less than 10 JTUs of turbidity, work reasonably well when the turbidity of the incoming water is 10–20 JTUs, and under occasional emergency conditions can be operated for a short time to produce an acceptable filtrate when the settled water has a turbidity of 20–50 JTUs. (Beyond this the whole works would go out of operation.)

Pressure filters have many of the characteristics of the rapid gravity type but are enclosed in steel pressure vessels and are normally used where hydraulic conditions in the system make their adoption desirable. They can be installed, for instance, at any point in a pressure pipeline without unduly interfering with the hydraulic gradient, and often eliminate the need for double pumping. While equally dependent on coagulants for their action, they do not commonly follow settling basins, which tend to break hydraulic gradients anyway, thus favouring the use of rapid gravity filters.

The mixed media filter[1] is a refinement of the rapid gravity sand type. Instead of a bed of sand supported on gravel with particles of somewhat similar density but greater size, various layers consisting of media of different densities are used. As a result, a very coarse upper layer of lightweight material (anthracite or pumice) can provide increased void space to store the impurities removed from the incoming water. Under normal operating conditions, with the turbidity of the incoming settled water < 5 JTUs, the performance is better and the filter runs are longer than for the rapid gravity sand filter. Incoming water of 30–50 JTUs turbidity can be treated in an emergency. The better performance is offset by additional capital cost.

The 'Iraq' type filter[2] was developed for small town use in Iraq in 1954. It is basically a rapid gravity sand filter of very simple construction, in which all controls are eliminated and automatic operation is attained by the simple expedient of building the pure water tank with its top water level to coincide with that in the filter. It is extremely cheap and effective but does not easily lend itself to 'scaling up' in larger installations.

Greenleaf filters are basically rapid gravity filters which are controlled on a rather unusual principle. Instead of the flow being kept uniform by an automatic controller, the outlets are controlled by weirs and the additional head required to maintain uniform flow is supplied by a rise in water level above the sand. These hydraulic conditions are not dissimilar to those of the Iraq type filter in reverse.

The slow sand filter is the old original type of filter and has a successful history going back about 160 years. It works without coagulants and is often found on reservoir- or lake-derived supplies. Its filtrate is of excellent quality, but its use has declined because it takes up a lot of space and is labour-intensive to operate. It will work with incoming water of up to 30 JTUs turbidity. Slow sand filters are rarely found working with settling basins as the two work best in entirely different circumstances. A weakness is that they are not effective in removing colour. However, these filters have considerable merit and in countries where land and labour is cheap and chemicals are dear the possibility of using them should not be overlooked. It is usually advantageous to install fine strainers before slow sand filters, particularly where outbursts of algae may be expected.

Diatomite filters are not commonly found on waterworks. They are compact, high efficiency filters which are suitable for armies in the field, and swimming pools, and for meeting short term emergencies. They are small and portable and depend on the deposition of filter powders of diatomaceous earth (kieselguhr) on porous filter 'candles' for their filtering action. They cannot deal with highly turbid water and because of extremely high head losses in the filter their running costs are high. For most practical waterworks applications they would probably prove to be less satisfactory than other methods.

RAPID SAND FILTERS

Rapid gravity sand filters

Flow of water through sand. The depth of water above a filter bed is generally about 2 m. The flow of water through sand is streamline flow and the loss of head is proportional to the velocity. The loss of head is also affected by the absolute viscosity of the water and the porosity ratio of the sand, and therefore temperature and the cleanliness of the filter sand both affect performance.

Filter sand has a grain size of 0.4–1.0 mm, which is large in comparison with many of the smaller particles carried by water (Table 9.1), and the rapid gravity sand filter can only operate as intended when coagulants have been used. Interception of the suspended matter by the sand is a fairly complex process, but most of the suspended matter is removed by adhesion to the surface of the sand grains. As the filter becomes dirtier the pores diminish in size, the

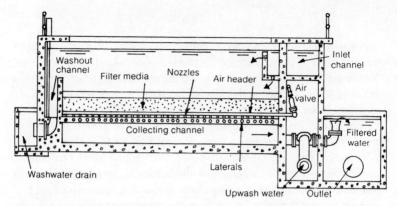

Fig. 9.1. *Rapid gravity filter (control valves not shown)*

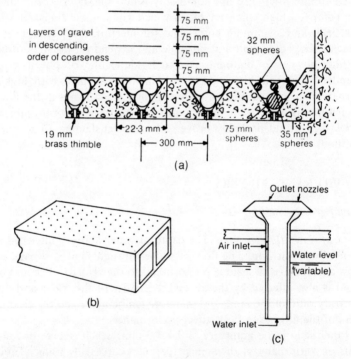

Fig. 9.2. *Filter bottoms: (a) Wheeler bottom; all spheres are of glazed earthenware; (b) perforated block type; glazed tile block used in the Leopold filter; (c) nozzle; the air depresses the water level in the lateral and escapes through the top inlet; when the air is turned off the water enters at both inlets*

Table 9.1. Relative size of sand grains and suspended matter

Material	Particle diameter (approx.), μm
Sand	500
Soil	1–100
Bacteria	0.3–3
Viruses	0.005–0.01
Floc particles	100–2000

velocity in the waterways increases and suspended matter is carried deeper into the filter. The filtering action therefore proceeds in depth and is only to a limited extent influenced by the formation of a film on the surface of the sand.

If the filter is working properly the sticky floc particles will not find their way through the intricate pattern of channels between the sand grains and perfectly clear water should emerge.

In the early part of a filter run when the sand is clean the floc particles are intercepted high in the sand bed, but as the top sand gets dirtier the floc particles penetrate more deeply and the loss of head increases. If the head loss in the sand at any point exceeds the static head of water on the filter, a vacuum will be induced which may cause dissolved air to be given off from the water and result in air binding of the filter. This occasionally happens just before the filter needs washing but disappears during the washing process.

Another phenomenon of dirty filters with high head loss is shrinkage of the sand surface and the formation of cracks. Surface cracking can be cured by paying greater attention to washing. The use of coarser sand is less desirable as turbidity of the filtrate occurs at lower loss of head than if finer sand is being used.

If through negligence a filter bed is allowed to become too dirty, the head loss becomes excessive and 'break-through' may occur. In this condition turbid water will pass to supply.

Construction. A rapid gravity filter (Fig. 9.1) is basically a bed of sand supported on a bed of gravel, embedded in which there is a system of underdrains. The whole is enclosed is an open-topped concrete chamber. The sand layer normally has a thickness of 0.45–0.75 m. The sand grains are fairly uniform in size, lying between 0.4 mm and 1.2 mm. About 10 % will pass an aperture of 0.45 mm and about 60 %

will pass an aperture of 0.65 mm. The coefficient of uniformity normally varies between 1.2 and 1.7 and the sand should be sharp, hard, clean and siliceous.

The supporting gravel beds may be omitted if porous plate filter bottoms are provided. Where gravel beds are required, they should have a total depth of about 45 cm, making the total thickness of the filter media about 1 m or a little more. The gravel should be graded in size between 2 mm and 60 mm.

Filter bottoms (collectors, nozzles, porous plates etc.) and the general underdrain system are normally supplied by specialist manufacturers and differ widely in detail. The basic requirement is that they should consist of a large number of orifices, uniformly arranged to collect filtrate from, and deliver air and washwater to, the filter bed (Fig. 9.2). At the same time they should preclude the passage of the filter media. The whole assembly must be carefully installed and dead level. The air scour will find escape through any high spots and give an uneven wash.

Each filter has four valves to control inlet, outlet, washwater and the drain to waste. If air is used to accompany the washwater a fifth valve on the air pipe is also required.

Filters may have areas up to about 100 m^2; above this, limitations are imposed by the rates at which air and washwater have to be provided and the distances the washwater (having passed upward through the sand) has to travel sideways to the drains. Additional filter capacity above 100 m^2 is provided by multiplication of the filter units; there is no limit to the number of units which can be installed.

Filter rates and performance. In recent years the rate at which the water is passed through the sand has been increased from about 3.9 m/h to 7–10 m/h. Little loss of efficiency has been noted but, obviously, the more sediment there is deposited in the sand in a given time, the more frequently the filters have to be washed. In swimming pool installations the rates may be as high as 12–15 m/h. Much depends on local conditions. High temperatures and slightly coarser sand grains facilitate the higher rates, and filters can in theory work with rates as high as 50 m/h (in practice they never do).

While it is common practice to state the working rate of filters in terms of flow through a unit of surface area in a given time (e.g., m/h), it should be remembered that filter performance depends essentially on adherence of the impurities to the surface of the sand grains themselves. The smaller the grain size of sand in any given container, the greater is the total surface area of all the sand grains. It therefore follows that the depth of the sand bed and fineness of grain contribute as much to filter efficiency as filter area.

The head loss through the filters should never exceed 2 m and is often less.

Filter control and washing. Operation at a prescribed rate of flow is ensured by fitting some form of controller on the outlet pipe or (less frequently) by varying the available head of water on the filter.

When operating normally the inlet and outlet valves should be open and the washwater and drain valves closed. When backwashing the two latter valves should be open and the inlet and outlet valves closed.

Filters may run for several days without washing, but on most manually controlled works they are washed daily by the day shift and allowed to run until the following day. The washing of a filter takes about 20 min and the units in the battery are washed in turn. In most plants the washing of one of a battery of filters may deflect additional flow through the other units. This can be tolerated because rates of filtration within reasonable limits are rarely critical. The washing operation lends itself to automation, and an increasing number of plants can be set, at predetermined times or loss of head, to shut down, wash and return each filter in a battery to service without manual attention.

The washing process might be by water alone or by air followed by water. The action of air is particularly effective in breaking up the crust which tends to form on the sand surface. The use of air scour is typical of European practice. In North America the crust is broken by surface water jets and rakes as well as by air.

If a filter is not washed thoroughly, the surface crust may crack and fragments might penetrate deep into the sand and form mudballs which are difficult to wash out. Signs of distress in a filter are surface cracks, the tendency for the sand to shrink away from the concrete walls, and non-uniform surface turbulence during the washing process. One of the merits of rapid gravity filters over the enclosed pressure type is that these signs can clearly be seen. One of the merits of manual control over automation is that the attendant has to be present and can observe what is happening.

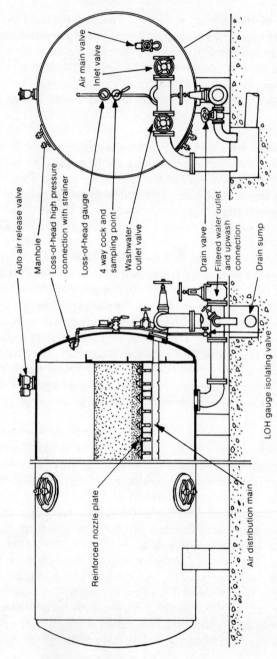

Air main valve

Inlet valve

Auto air release valve

Manhole

Loss-ol-head high pressure connection with strainer

Loss-ol-head gauge

4 way cock and sampling point

Washwater outlet valve

Drain valve

Filtered water outlet and upwash connection

Drain sump

Reinforced nozzle plate

Air distribution main

LOH gauge isolating valve

Fig. 9.3. Horizontal air-scoured pressure sand filter (by courtesy of Paterson Candy International Ltd)

At the prescribed head loss, or time, the filter is taken out of service (manually or automatically) by closing the inlet and outlet valves. The drain valve is then opened to lower the level of water to the level of the weir wall that holds the sand bed in position. The air scour is operated for about 3 min. This forces air upwards through the sand at a rate of about $1-1.5$ m^3/min per m^2 under a pressure of 0.42 kg/cm^2. When the water over the bed has become thoroughly agitated, upwash water is turned on at a rate of about 0.5 m^3/min per m^2 and allowed to run until it clears and the sand bed can be seen. In practice this will be for about 5 min, during which time the washwater will be falling over the cross-wall weir and escaping through the open drain valve. In some filters air scour and water scour can be applied simultaneously.

It is during the application of washwater that the maximum practical dimensions of the filter assume importance, as 50 m^3 of water per minute to be transported to, up and across and then away from a filter is quite a lot of water to move about. While this washwater is flowing the sand bed expands, lifting the surface. The weir over which the washwater escapes has to be higher than the sand surface when expanded (or say at least about 40 cm higher than the bed when at rest) or there will be loss of sand. The pressure of the washwater at the nozzle outlets should not exceed 0.6 kg/cm^2. The washwater is designed to rise upwards at about 4–5 times the rate of normal downward filtration. The amount of washwater used should average about 1.5% of the total daily throughput, with some variation up or down to accord with the condition of the incoming water.

When the filter has been washed, the drain, upwash and air valves should be closed (if this has not already been done) and the inlet valve opened to recharge the filter. Return to service is generally effected by opening the outlet valve slowly to give the filter bed time to settle down and rid itself of any loose sediment stirred up during the washing process. Some filters have slow-start valves fitted which restrict their capacity to work up to full speed for half an hour or so.

Pressure filters

There is no theoretical difference between the operation of a rapid gravity filter and one of the pressure type. Rates of flow, criteria for washing and most other factors remain similar. Instead of the filter assembly being housed in an open concrete box it is enclosed in a steel pressure cylinder which permits it to be installed in pipelines without breaking the hydraulic gradient and thus introducing the need for double pumping. In most cases, however, open rapid gravity filters are preferred on the grounds that it is possible to see what is going on, but above all because of the greater ease of regulating the flow, mixing the coagulants, and pre-treating the water. In a typical pressure filter installation there is a noticeable absence of pre-treatment and flow controllers and as these have considerable value it follows that pressure filters must be working at a disadvantage. A trip round a pressure filter plant, however, has the merit of applying a slight corrective to the advocates of excessive pre-treatment and ultra-refined gadgetry because most pressure filter plants appear to be functioning quite well with a minimum of external equipment.

The filter shells are commonly of 2.4 m diameter and can be installed vertically or horizontally. Horizontal units (Fig. 9.3) are normally encountered in large installations where the small sand area exposed in a vertical shell (5 m^2) would necessitate too many units. The length of a horizontal unit does not normally exceed about 15 m and thus the unit has 2.4 $\times$ 15 m^2 = 36 m^2 of sand area. Theoretically the round shell of a horizontal filter constricts the disposition of the underdrains and detracts from the efficiency of the filter, but this does not seem to matter very much in practice.

Where reservoir water, which is naturally reasonably clear, is transported in long, fairly steeply inclined pipelines (conditions commonly found in the north of England), pressure filters are common. The water can be filtered without preliminary treatment in settling tanks and without breaking the hydraulic gradient.

In other places, however, pressure filters are often used only on smaller works or where simplicity of operation might be advantageous. There is a noticeable tendency to keep the designed filter rates rather low (4–6 m/h). However, much of this conservatism may be due to prejudice, and where the pressure filter can usefully be installed its adoption can be recommended with confidence.

Multi-layer filters

The action of washing a sand filter disturbs the graded beds of sand and gravel, and the capacity of the filter to regain its designed sequence of layers after washing is due to the natural capacity of the coarser gravels to fall through water more quickly than the finer sands. No matter how many times the filter is washed the sand layer will remain on top. Because of this characteristic, the filter does not follow the natural order of water treatment procedure. In all other processes the bigger impurities are removed first. The filter, with its finest layer at the surface, removes all impurities at one fell swoop.

This is undesirable because not only does the sand do all the work but the void spaces in the sand are relatively small, there is limited capacity to store the entrapped impurities and therefore the filter rapidly gets dirty.

There are two ways to overcome this problem:

(a) by reversing the flow through the filter so that the water passes upwards, deposits its heaviest impurities in the underlying gravel and uses the upper layer of sand to catch only the finest particles;

(b) by placing a coarse layer of very lightweight material on top of the sand in a normal downward-flow filter, to stop and store most of the particles before they reach the sand; the materials commonly used are anthracite or pumice which are so light in comparison with sand that they sink more slowly after washing and remain on top.

Either method removes the bulk of the floc before it reaches the sand and stores it in the large voids. As a result more turbid incoming water (even up to 50 JTUs) can be treated, and longer filter runs using less washwater can be obtained. (Like all others, multilayer filters work best when the turbidity of the incoming water is < 5 JTUs.)

Over the years, method (b) appears to have been the more widely practised. There are comparatively few upward-flow filters about, possibly because they are not too easy to wash. In an upflow filter the method of washing is to accelerate the upflow to drive the dirt from the gravel right up through the overlying sand. There would appear to be little amiss with the basic principle but the idea has not enjoyed widespread popularity. In some designs excessive expansion of the filter bed during washing is restrained by a metallic grid on the surface of the sand.

Perhaps the reason why upflow filters have not caught on is that method (b) has proved completely effective. In UK practice anthracite (1.25–2.5 mm) is used. In Switzerland pumice is favoured.

Because of their exceptionally high performance, it is possible to use multi-layer filters without settling basins where turbidities are not too high. The filters are also very effective where algal growth is feared because of the large capacity for storage in the void spaces of the upper layer.

The use of these filters is growing as they produce a high quality filtrate and offer the advantages mentioned. However, they are expensive in first cost and trouble has been experienced in the past because the lightweight materials have tended to be rubbed down by abrasion during the washing process and have needed to be replaced more often than sand. Where the material has to be imported this might be a consideration which militates against the adoption of multi-layer filters, and perhaps they will not make undue progress in developing countries within the foreseeable future as a result. In technically advanced countries multi-layer filters are becoming very common.

It is clearly possible to add a layer of anthracite to existing rapid gravity sand installations, where the carrying capacity of the under-drains and outlet pipes is big enough to make such conversions feasible.

A development in the USA[1] is to increase the number of layers to three: 10% high density garnet sand, 30% silica sand and 60% anthracite. However, there is a tendency for the layers to intermingle. Filtration rates up to 24 m/h are claimed, but most installations appear to be working at no more than half this rate.

'Iraq' type filters

About 1954 Iraq embarked on a programme for upgrading the waterworks in the smaller towns. The delay and expense of importing equipment from abroad resulted in the design of a simple form of treatment plant, most of which could be built cheaply using local materials.[2] The three basic elements of the works were the settling basin, a rapid gravity filter and a high level tank.

The settling basin was of the standard horizontal-flow type which had proved eminently suitable for the silty Iraqi rivers, and the rapid gravity filter in no way departed from standard except that it was locally produced. The high level tank, which could be built in concrete or from bolted steel plates, was not remarkable except that the somewhat tricky foundation problems commonly found in southern Iraq could be solved by standing the tank on the concrete treatment unit. The ground in southern Iraq will normally bear only about 5–7 t/m², but this is more than sufficient to carry the very low foundation loads imposed by typical waterworks structures. The surplus load-bearing capacity of the ground could be usefully employed, therefore, by standing the high level tank on the treatment structure, which could be designed to act as a raft footing.

Pumps, pipes and valves were readily available in the local market, and the only pieces of technical equipment needed were the rate-of-flow controllers and the chlorinator. The cost of importing these was not prohibitive, but cheaper alternatives could be found. It was noted that if the pure water tank was raised from its usual level below ground so that the top water level coincided with top water level in the filters, three extremely desirable results could be attained for nothing. In an imported filter, these could be provided only by spending money in one way or another. The advantages were as follows.

(a) Flow control was automatically provided because the increased loss of head due to the filter becoming dirtier was balanced by a fall of water level in the pure water tank, thus dispensing with the need for a flow controlling device.

(b) Constant output was ensured by the high lift pump. No matter what state the filter was in, the speed of operation had to match that of the pump, which remained constant.

(c) In any mechanical flow controller there is an imposed head loss that declines as the natural losses in the filter increase. No such artificial head loss occurs in the Iraq type of filter so there is a gain in efficiency.

The savings in cost and time needed for construction were remarkable. A few of these plants were built in Iraq at that time and have been copied to some extent in other Eastern countries, but they are not as widely known as they might be, probably because they offer no scope for commercial exploitation. As no essential item in their construction has been skimped or omitted it is difficult at first glance to see why they can be built for less than half the cost of an imported works. The explanation lies only partially in the omission of all frills and gadgets, which, besides adding to the initial cost, tend to increase subsequent maintenance problems. The most significant saving is made because the basic waterworks materials—pumps, valves, meters, pipes etc.—can be purchased at local market

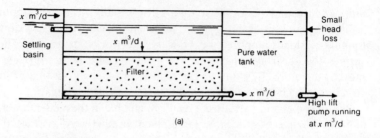

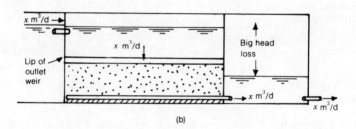

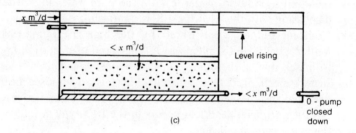

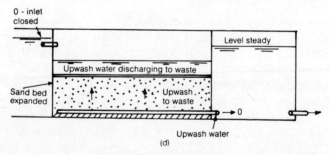

Fig. 9.4. Iraq type filter: (a) running with filter clean; (b) running with filter dirty; (c) shut down in preparation for washing; (d) during washing

rates without having to bear heavy on-costs to cover the design, import and marketing expenses inevitably incurred in a complete package deal.

Plant details. The usual care has to be taken about adequate mixing of the incoming chemicals and the provision of flocculation where necessary.

The inlet valve between the outlet channel of the settling basin and the filter could be of the hand flushing type as used on sewer flushing chambers, which gives swift and simple action. The drain valve could best be a spindle-operated penstock. The filter outlet valve and the washwater valve would have to be ordinary waterworks pattern valves, of the smallest diameter possible to permit quick opening and closure. Valves should not work with velocities in excess of 3 m/s, which would be the critical design factor.

If the underdrain system were locally constructed with asbestos-cement laterals, the air nozzles could be brass plugs drilled with 3 mm orifices on the top of the laterals and the water inlets provided by 13 mm holes on the bottom. The laterals are dead level and so they can be used for air and water intermittently, as the air when admitted runs along the crowns of the pipes and gives a uniform wash. The air holes at Bagdad were of $\frac{1}{8}$ in dia. spaced 2 in apart on the crown of the laterals, the water holes of $\frac{1}{2}$ in dia. spaced 5 in apart. The laterals themselves were $2\frac{1}{2}$ in dia. copper tubes at 12 in centres. The holes were drilled straight into the copper tube.

The total area of the 13 mm dia. water orifices should be not less than 0.3% of the surface area of the sand; velocities in the laterals should be not more than 0.5 m/s, and in the main header pipe not more than 0.8 m/s. These are filtering velocities, which would be exceeded during the washing process.

The low lift pump (from river to settling basin) should have an output slightly in excess of the high lift pump, so that the plant is able to operate at its design speed at all times with a slight surplus of water escaping over the basin overflows.

The air pipe must at some point in its passage from blower to filter be made to rise above water level, and must be fitted with air and reflux valves to prevent back-siphonage, which might damage the air-blowers.

Normal operation. For normal operation, the low lift pumps are started and all dosing gear is switched on. Care should be taken to see that the water level in settling basin, filter and pure water tank is at a constant high level before the high lift pump is started. When the filter is working the inlet and outlet valves should be open, and the upwash, air and drain valves closed. The only resistance to flow through the unit is the filter sand and as this is clean the initial loss of head, reflected by a drop in level in the pure water tank, may be about 15–20 cm. When the water in the pure water tank has fallen to this level the flow pattern stabilizes: the flow into the filter, the flow out of the filter, the flow out of the pure water tank and the pump output are all equal. The situation is as shown in part (a) of Fig. 9.4.

As the filter gets dirtier, the flow pattern cannot change but the resistance of the sand gets bigger and the level in the pure water tank falls (part (b) of Fig. 9.4).

The washing cycle is shown in parts (c) and (d) of Fig. 9.4. When a preselected loss of head is reached the high lift pump can be automatically tripped out and the level in the pure water tank will rise. After a brief period the low lift pump can be stopped. Inlet and outlet valves are closed, and the drain valve is opened. When the level in the filter has fallen to outlet weir level, air and upwash water are admitted and the filter is washed normally. The upwash, air and drain valves are opened, and all other valves are closed during this operation.

Finally drain and upwash valves are closed and the filter returns to the starting condition preparatory to another filter run.

Limitations. The treatment plants built on the Iraq principle were mainly rather small. The plant at Shalchiya, Bagdad, had a designed output capacity of 11.5 mgd (about 50 000 m³/day) and one at Ulu Selangor South in Malaysia about 15 000 m³/day. There are several others, but all are smaller. The difficulty of scaling-up the design is that it is considered undesirable to have one high lift pump controlling more than two filter beds. Assuming the maximum area of one filter bed to be 100 m², it follows that the maximum output of a twin filter unit would be about $2 \times 100 \times 23 \times 7$ m³/day $=$ 32 200 m³/day.

The filters at Shalchiya were smaller than the maximum possible size; they worked at about 5 m/h, and for the 11.5 mgd three parallel units were installed.

To increase the size of works of this type beyond about 30 000

m^3/day, construction of parallel units becomes essential and once this happens the benefits of scale diminish. However, the large cost differential is such that no conventional works could ever be as cheap. The sedimentation basins and filters are not affected; only the pure water tanks and pumps have to be duplicated or triplicated as circumstances demand. The total pure water tank capacity and the total pump horsepower are not altered.

The size of the pure water tank does not affect the principle on which these plants work, but a bigger tank has the slight advantage of providing a 'built-in' slow start after the plant has been washed. The merit of starting a filter at reduced rate after washing is probably exaggerated, however. There is no doubt that it is beneficial, but there are many works where the practice falls somewhat short of the ideal.

Declining rate filters

There are two ways in which rapid gravity sand filters can be categorized:

   (*a*) constant rate, with constant or variable head loss;
   (*b*) declining rate, with constant or variable head loss.

Prior to 1960, constant-rate, constant-head-loss filters, as described on pages 113–122, were almost universally used, the constant rate being achieved by some sort of flow controller on the outlet which opened as the sand grew dirtier to maintain a constant overall head loss. These controllers were of many varieties but venturi-controlled butterfly valves were common.

The 'Iraq' type filter also operates at constant flow (as dictated by the high-lift pump) and constant overall head automatically provided by a declining level in the pure-water tank.

Declining rate filters are similar in principle to the Iraq type, in that the increasing difference between the water levels within the filter and at the outlet overflow weir automatically compensates for the increasing resistance in the sand. However, this difference is attained by making it possible for the water level over the sand to increase while the outlet level (weir-controlled) remains constant. The merit over the Iraq type is that the whole battery of filters is discharging at the same downstream head and is controllable by one pump or one main outlet weir.

It is standard practice to wash declining rate filters on a pre-

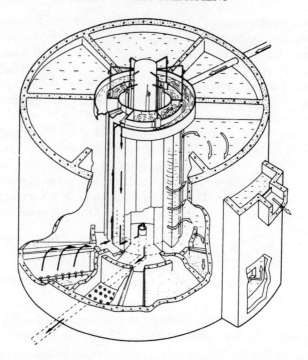

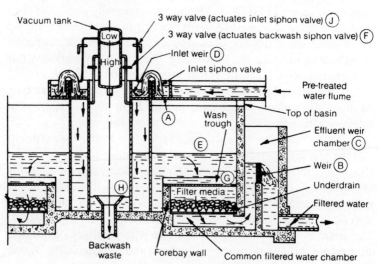

Vacuum tank

Low

High

3 way valve (actuates inlet siphon valve) (J)

3 way valve (actuates backwash siphon valve) (F)

Inlet weir (D)

Inlet siphon valve

Pre-treated water flume

Top of basin

(A)

Wash trough

Effluent weir chamber (C)

Weir (B)

(E)

(G)

Underdrain

(H)

Filter media

Filtered water

Backwash waste

Forebay wall

Common filtered water chamber

Fig. 9.5. Greenleaf filter (by courtesy of Infilco Inc.)

selected time sequence, each filter being taken out of service at some given time in the operating cycle of 1–3 days, depending on the turbidity of the incoming settled water.

The inlet channel is generously sized so as to maintain a constant head (and therefore constant inlet flow) at all the filter inlet weirs. If and when the dirtiest filter cannot accept the full rate, the water level in the filter shell rises and drowns out its inlet weir. This progressively reduces the flow rate through the dirty filter and, as the total incoming flow is constant, raises the level of water in the inlet channel, thus causing an increased flow through the other filters by equal distribution. This effect becomes even more marked when the dirty filter is taken out of service for washing, but it is not too serious in practice because the shut-down takes place on a filter which has already declined in output.

In recent years quite a number of new filter plants have been designed on the declining rate principle, and in Korea, for instance, these filters have become universal.

Greenleaf filter

The Greenleaf filter (Fig. 9.5) is a rapid gravity sand filter of fairly conventional type except that it has a control system of unusual design. These filters are widely used in the USA and Japan and are also found elsewhere, particularly in SE Asia.

The rising loss of head in the filter is balanced by the rising level of water above the filter sand (as opposed to the falling level in the Iraq filter's pure water tank). Washing is effected by creating a vacuum in the backwash siphon. There is no air scour.

The principle is quite sound, and choice must depend purely on economic grounds. Filters may be round, square or rectangular to suit design circumstances.

Operation. For the initial start-up, water from the pre-treatment unit enters the control centre and passes to a distribution channel (Fig. 9.5). The inlet siphon valve is opened, and water passes via the inlet weir chamber, fills the forebay and enters the filter cell proper. It then passes normally through the media and the underdrains, a positive head being maintained on the filter by weir B.

During operation with a clean filter, the filter cells fill to a height slightly above the outlet weir B. The rate of flow to each filter unit is controlled by setting the inlet weir D to the required level.

As filtration continues the head loss in the media increases and the water level E is forced to rise to compensate. The rate of flow is still controlled at the preselected level by D. The filters can all operate at their required speed irrespective of differing water levels in the individual cells.

When maximum head loss has been reached, water from the inlet cell spills over into the backwash siphon weir (not shown in Fig. 9.5) and the inlet siphon valve can be manually or automatically vented, thus stopping flow to the filter. As inflow stops, the level of water in the filter falls to that of weir B.

The backwash sequence commences with the opening of the three-way valve F, which connects to the high vacuum tank. Water is drawn from the forebay up the legs of the backwash siphon, and pulled upwards through the filter media from the filtered water chamber. As soon as both legs of the backwash siphon are primed, the backwash continues by siphonic action and water leaves the cell unit down the backwash waste pipe. Washwater is provided by the other filter units which feed into the filtered water chamber. The difference in water levels between B and G provides the upwash action, and the difference between G and H provides the head to operate the backwash siphon. No other controls are necessary.

The backwash cycle is terminated by venting the backwash siphon through valve F, and water level E rises to coincide with B. Return to service is brought about by opening inlet siphon valve J, and conditions return to those of initial start-up.

These filters are simple to operate, with ease of control afforded by the siphons, the comparative absence of flow controllers, and the absence of washwater tanks and pumps.

SLOW SAND FILTERS

Slow sand filters were first used in London in 1820 and there are still many big cities (including London) where they are preferred to the rapid gravity type. There is a tendency nowadays, however, to consider them as being outmoded and to assume that rapid filters should be adopted. This is not strictly true: slow sand filters are still being built, the resulting filtrate is of very high quality and it is unlikely that they will be replaced in areas where they are already installed.

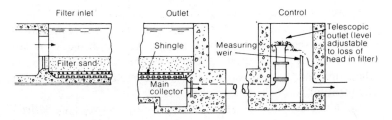

Filter inlet   Outlet   Control

Telescopic
outlet (level
adjustable
to loss of
head in filter)

Shingle   Measuring
weir

Filter sand

Main
collector

Fig. 9.6. Slow sand filter

They are of simple construction (Fig. 9.6). A bed of sand some
0.9 m deep rests on graded gravel in which underdrains of open-
jointed tiles are buried. The sand is carried to the full depth of the
bed near the outer walls, and to avoid carrying it into the underdrains no pipes should be laid within 0.6 m of the walls. The sand is
normally finer than that in a rapid filter, and its quality and grading
are less exacting, (Table 9.2), so it is more likely to be found locally.

Because cleaning is a lengthy process not less than three filter
beds should be available, as one is frequently out of service for
cleaning.

No pre-treatment or chemicals are absolutely necessary, although
strainers are sometimes installed. Water with turbidities of up to
30 JTUs can be put directly on to the filters at rates up to 5 m/day.
(Such turbidities are higher than for most reservoir-derived waters,
for which slow sand filters are specially suited.)

The merits of slow sand filters are that the quality of the filtrate
is high and all harmful bacteria are removed; no chemicals are
required or even desirable; the filters can be built with local materials
and local sand can often be used; and operation is easy.

The disadvantages are that they cover a large area and building

Table 9.2. Typical slow sand filter construction

	Depth, m	Grading, mm
Water	1.2–1.8	—
Sand	0.7–0.9	0.2–0.4
Fine gravel	0.05	5–10
Medium gravel	0.05	10–25
Coarse gravel	0.15	25–80
Underdrains	—	—

costs are high; they do not deal effectively with highly turbid water, nor will they remove colour; algal growths cause operational trouble; and the filters are labour-intensive. As regards the last point, some of the washing procedures can be mechanized, and a need for labour may not be a major disadvantage where the labour is cheap and plentiful.

The filters are very suitable for use on reservoir- or lake-derived supplies for small communities where technical supervision is lacking.

Operation

The principle of operation of slow sand filters is different from that of rapid filters, the main difference being that the filtering action mostly takes place at or near the surface of the sand. Here a mat (or schmützdecke) rapidly builds up, the filter becomes 'ripe', and filtration continues for a period of about a month before cleaning. As the rate of flow is low the loss of head is small, varying between about 5 cm when the filter is clean to about 90 cm when it is dirty. The increased head required is provided by adjusting the outlets so as to raise the head available over the sand.

When the filter is losing about 90 cm of head it is taken out of service and drained down, and the top 15 mm of sand is scraped off and removed to be washed. The filter is then returned to service slowly in order to permit the schmützdecke to re-form.

After several scrapings the sand bed becomes too thin (about 70 cm) and it has to be restored to its original thickness (90 cm) by replacment of the sand removed. This is generally the same sand that was removed previously, which has been washed and stored in the vicinity of the filter.

After a period of several years, the bottom sand in the filter tends to turn black and septic and the bed may need to be dug up and reformed.

Use after strainers

The performance of slow sand filters can be materially improved if the raw water is passed through strainers before being admitted to the filters. These commonly take the form of rapid gravity sand filters, as installed on London's water supply at Kempton Park and at the Ravensthorpe works at Northampton. These filters run without coagulants and act purely as high speed strainers; they are particularly effective in removing algal growth. An important secondary function that they can perform is to control tastes where the water has been dosed with activated carbon; the strainers are used to remove the powder before it reaches the slow sand filters.

Microstrainers might also be used where conditions suit them; i.e., where the water is fairly clear but prone to outbreaks of algae.

It is normally possible to work the slow sand filters faster when preceded by strainers. At Ravensthorpe the output of the old filters was doubled after the rapid gravity plant was installed. The cost of the rapid gravity plant was much less than the cost that would have been incurred if the slow sand beds had been duplicated.

WASHWATER TANKS AND PUMPS

In most conventional filter plants a washwater system must be provided. The water used for backwashing the filters must itself have been filtered and in most cases it will have been fully treated. It is usual for filters to be washed one at a time so that the total rate of flow through the works suffers the minimum of disturbance. The washwater flow has a rate about five times the normal filter rate and lasts for about 7 min, and therefore the total water required per wash is about $(7/60) \times 5 \times$ normal hourly filter rate $\times$ area of the filter. As the required rate of flow of washwater, and thus the required capacity of pipework, pumps and washwater equipment, depend directly on the area of the filter, this imposes a limit on the filter size.

The pressure of the washwater required at the filter nozzles is about 0.6 bar, which can be provided by washwater tanks situated on the filter house roof or by special washwater pumps. If tanks are installed they provide storage and the size of the pumps feeding them can be reduced, as there is no longer a necessity to match pump capacity with upwash rate.

The washwater tanks should never have a capacity less than that required for one complete filter wash, and in practice they are normally made somewhat bigger to obviate periods of waiting for the tank to refill when the whole battery of filters is being washed, as is customary, by the day shift. The maximum amount of tankage required should never exceed the equivalent of about 30 min total station output on a small works, and as the works get bigger it gets proportionally smaller (in terms of total output) because there are more filter units. The top water level of the washwater tanks is normally about 9 m above the filter bed level.

It is not considered desirable to use high pressure water from the pumping main for filter washing. For one thing the pressure would need to be reduced, and if this were done continuously over the life of the plant it would represent a significant waste of energy.

PURE WATER TANKS

When the water leaves the filters it passes into a pure water tank before being pumped, or gravitating, to supply. These pure water tanks serve three purposes. They maintain output for short periods when the filters are shut down, they provide balance between the flow rates of the filter and the high lift pumps, and incidentally they provide adequate contact time for the chlorination process. (Because the pure water tanks act in this way, specially provided chlorine contact tanks are rarely necessary.)

Pure water tanks are commonly made with a capacity equivalent to 2 h of works output. There is little logical basis for this but widespread custom suggests that the practice has proved satisfactory.

In the Iraq type of plant, where the pure water tanks are central to the whole method of operation, they can with advantage be made much smaller. It is true that smaller tanks reduce the 'slow start' effect which is provided by these treatment units, but in practice this disadvantage does not seem to matter very much.

In pressure filter installations, pure water tanks are often omitted because the considerations that dictate the use of pressure filters do not favour breaking the hydraulic gradient for the purpose of inserting a tank.

In the Greenleaf type of filter the effluent weir chamber can be increased in size and serve as a clear water well (or pure water tank— the terms have much the same meaning).

10: Disinfection

The final process in water engineering is disinfection. The widespread use of rapid sand filters results in a final filtrate of great clarity which is, however, not always free of bacteria and other organisms. These have to be reduced, either completely, or certainly to negligible proportions, by some form of disinfection. This can be done by the addition of certain chemicals, by ozone, by ultraviolet light or by boiling. In practice the vast majority of waterworks use chlorine or chlorine compounds.

The excess lime process of softening kills bacteria. It is not widely used but because of the bactericidal effect of the high pH values incidental to the process, it is occasionally adopted where the raw water is not only hard but also polluted.

Ozonization is fairly common, particularly in France and Switzerland, and produces excellent water of high quality. The process depends on the production on site of ozone, O_3, by the passage of high tension, high frequency electric discharges through the atmosphere. The ozone is subsequently absorbed in the water to be treated and has a powerful bactericidal action without imparting taste to the water. It is a costly process needing skilled maintenance and may not yet be considered suitable for adoption in developing countries in preference to chlorination. In addition to its high cost it suffers from the fact that ozone is not very persistent and water treated by this method at source might be reinfected at a later stage in the distribution system.

Ultraviolet radiation is little used outside Russia. Boiling is of value to consumers in emergency but cannot be applied on a plant scale in normal practice. The various alternative substances (e.g., silver, permanganate, bromine, iodine) do not compare in cost or effectiveness with chlorine and may be ignored.

CHLORINE AND CHLORAMINES

Bleaching powder (chlorinated lime) and sodium hypochlorites of different commercial brands are widely used for sterilizing small water supplies. Chloride of lime is a powder and is sold in drums. It is unstable and when the drum has been opened it loses its chlorine strength fairly rapidly. A freshly opened drum has a chlorine strength of about 33% by weight and the dosage can be calculated accordingly. The lime content is insoluble and a solution should be mixed and allowed to settle before dosing. The solution also loses strength and batches should be small and used up within a few days.

Sodium hypochlorite solution is a clear liquid which can be fed through solution-feed equipment without fear of clogging. It can be bought in various strengths and is normally diluted to a 1% solution before application. As with chloride of lime, it tends to lose strength if exposed too long before use. Injection is by gravity feed or by special chlorine pumps.

In clear water 0.5–1.0 mg/l of free residual chlorine should ensure sterilization. Doses of this magnitude are commonly used on the small, unsophisticated works for which this method is suitable.

Chlorine gas

In most modern works of any size, chlorine is supplied as a liquefied gas in cylinders or drums and injected into the water supply through a chlorinator. The chlorinater is a fairly complicated apparatus which reduces the pressure of the gas leaving the cylinders, controls the rate of flow, mixes the gas with water and delivers it to the pump or injector which forces it into the filtered water. Once wet, chlorine is very corrosive, and all piping and pumps have to be of suitably resistant materials.

Many directives and guidelines[1-3] have been issued. There is a drift towards vacuum systems, residual monitoring, and electrolytic chlorination in Europe.

If the rate at which the chlorine leaves the cylinder is too high, the rapid transition from liquid to gas causes extremely low temperature, which, in turn, causes crystals to form which block the feed pipe. The remedy is to lower the rate of flow leaving the cylinder and this can be done by increasing the number of cylinders connected to the manifold leading to the chlorinator. The critical rate of draw-off varies with the size of the cylinder. A 1 t drum can discharge at up to 10 kg/h in average temperatures but a 70 kg cylinder will 'freeze' at very much lower rates. There is no theoretical limit to the number of cylinders that can be connected in parallel, but in practice, in big works, where flows exceed 60 kg/h it becomes cheaper to insert a special 'evaporator' between the cylinder and the chlorinator. With an evaporator the chlorine leaves the drum in liquid form and 'freezing' does not occur.

It is not unusual to install a completely duplicated chlorine system in which two or more cylinders are provided in parallel banks with an automatic, pressure-controlled switch to change from one group to the other when one battery of cylinders is empty. In many countries the cylinders have to be owned by the waterworks. Sufficient cylinders should be held in store to replace an empty bank and there should be enough surplus cylinders to cover the time required to send them to the factory to be refilled. Assuming one bank of cylinders will keep the works going for a month, the minimum number required would be

(a) the number of cylinders in the working bank;
(b) the number of cylinders in the reserve bank (connected);
(c) a number equal to one bank in store (not connected);
(d) an equal number to those stored away being refilled.

If there were undue delay in getting cylinders refilled, item (d) would need to be increased.

Action of chlorine

The exact reason why chlorine kills bacteria is still debated but it is now believed that it is by direct toxic action. Until recently there was a theory that the bacteria were killed by oxidation as a result of the release of nascent oxygen, which occurs when chlorine is injected into water; this is no longer considered probable, although the oxygen released clearly has an effect on any algae, iron and manganese which might be present.

When added to water, chlorine reacts to form hypochlorous acid and hypochlorite:

$$Cl_2 + H_2O \rightleftharpoons HClO + H^+ + Cl^-$$

The hypochlorous acid and the hypochlorite together represent the 'free available chlorine', which is a very powerful bactericide. The hypochlorous acid is more powerful than the hypochlorite.

If ammonia is present, either naturally or because it has been added, chloramines will be formed as follows:

$$NH_4 + HClO \rightleftharpoons NH_2Cl + H + H_2O \quad \text{(monochloramine)}$$
$$NH_2Cl + HClO \rightleftharpoons NHCl_2 + H_2O \quad \text{(dichloramine)}$$
$$NHCl_2 + HClO \rightleftharpoons NCl_3 + H_2O \quad \text{(trichloramine)}$$

The relationship between the amounts of the three types of chloramine depends on the pH and the NH_4 concentration of the water. As trichloramine can only form at very low pH values, the other two prevail in water treatment, dichloramine being much the more powerful bactericide. The total quantity of dichloramine and monochloramine is referred to as 'combined available chlorine'.

Post-chlorination

The following comments refer only to post-chlorination of water which has already passed through the treatment plant.

Simple chlorination (break-point chlorination). Even the purest water tends to have slight traces of ammonia and other compounds. These immediately react with chlorine to form chloramines, which are themselves bactericidal. At one time this was considered to be desirable and ammonia was added to intensify the effect (see below). Although this practice is still fairly common, it has declined in popularity in favour of 'break-point' chlorination. This is because free chlorine is quicker in action and more powerful than chloramine.

Break-point chlorination simply means that the chlorine dose is added initially to oxidize any reducing compounds present, then increased to form chloramines with any ammonia present, increased still further to destroy the chloramines, and increased finally to build up the highly bactericidal free residual chlorine.

As it passes through the third stage (destruction of chloramines) there is a noticeable dip in the chlorine residual present and this is known as the break point. When this point is passed the free residual rises again, more or less in step with the chlorine being added, and a low free residual (low-tasting) chlorine will rapidly disinfect any clean water. The 'chlorine demand' of any given water is the amount of chlorine required to take the reaction through the break point. In some waters this is very low and often does not exceed or even reach 0.3 mg/l, but in other waters it may be much higher and has to be determined by experiment. All waters need a minimum of about 0.2 mg/l of free residual chlorine to effect sterilization (about twice as much at high pH values), and to obtain this it is unlikely that less than 0.5 mg/l of chlorine gas by weight would have to be added.

Free residual chlorine acts swiftly provided that the gas is well mixed into the water. Times of 10–20 min are often quoted but in fact, with the residuals of 0.5 mg/l normally used, 'kill' is generally much quicker and the time spent by pure water in trunk mains and various forms of storage tank almost invariably provides more than sufficient contact time. Although 'contact tanks' are frequently talked about, in practice they rarely need to be provided, because most works have pure water tanks which incidentally perform this function. It is a point that needs watching, however.

Rate of kill is adversely affected by low temperature and high pH. This latter effect is due to the formation at medium–low pH of hypochlorous acid in the break-point process and of dichloramines if ammonia is present. Both are most effective sterilizers in their respective spheres. The less effective hypochlorites and monochloramines form freely and predominate at higher pH, with resulting loss of efficiency. Low temperature adversely affects the action of chloramines but does not delay the action of free chlorine residuals to the same extent.

Ammonia chlorine process. The practice of adding ammonia to chlorinated water to achieve a NH_4:Cl ratio of 1:3 is still fairly common. It has the merit of being very persistent and maintaining a toxic dose far into the distribution system without creating taste problems to the same extent as does chlorine alone. However, it is less powerful than free chlorine and acts more slowly. At doses of 0.3 mg/l of chlorine and 0.1 mg/l of ammonia, contact time of about 20 min is essential, but as the time spent in pure water tanks, trunk mains and service reservoirs is normally to be measured in hours, the contact time available is more than adequate.

High ammonia doses are dangerous as they further slow down and nullify the toxic action of the chlorine, a consideration of particular concern in swimming pool control, where an unknown amount of ammonia may be introduced by the bathers themselves.

If it is required to maintain appreciable residual chlorine and chloramine to the furthest extremities of the pipe system, the initial doses injected at the works must be higher than those quoted above.

Super-chlorination and de-chlorination. Some waters are of varying quality, particularly deep well waters in fissured rocks which, while normally pure, might be subject to occasional gross pollution. To guard against this possibility, heavy doses of chlorine must be added continuously which, because the water habitually has little chlorine demand, makes the water unpalatable. The dechlorinating agent used is sulphur dioxide, which is supplied in cylinders and dosed into the water as a gas, using the same sort of apparatus (a chlorinator) as is used for chlorine. The final free residual left in the water can be regulated within close limits and is normally about 0.3 mg/l.

Pre-chlorination

As more fully described in Chapter 3, if a source of water is grossly polluted but not particularly turbid, or if algae have to be killed prior to removal in upward-flow settling basins, chlorine is added before the water passes into the treatment plant. Doses of 2–5 mg/l are normally used to counteract gross pollution, but most algae require only about 1 mg/l. In hot countries where sludge deposited in settling basins tends to go septic, a small dose of chlorine will control this. At Tehran 2 mg/l of chlorine is used purely to control septicity. Outbreaks of molluscs, bloodworms and slime formations in the works inlets require heavy slug dosing at high chlorine concentrations. When necessary this should be done while the treatment plant is not working. If the carrying capacity of raw water mains is reduced by slime formation, it can often be radically improved by the passage of slugs of water carrying chlorine doses of up to 50 mg/l of chlorine. This was done with marked effect on the Klang Gates main at Kuala Lumpur in 1964.

Control and safety measures

Chlorinators are very reliable but as a check on the amount of chlorine being used it is common practice to mount the gas cylinders on a weighing machine. The daily loss of weight of liquid chlorine can be compared with the amount used as indicated by the chlorinator setting and any discrepancy investigated.

Chlorine is a poisonous gas that will kill people if sufficiently concentrated, and the pipework should be tested for leakage at regular intervals. An open bottle of ammonia passed along the exposed pipes will give off a white vapour if chlorine is escaping.

Chlorine should be treated with respect, but it is not the most dangerous of deadly gases and in practice it gives very little trouble. The most likely accidents seem to arise from the cylinder outlet valves and if these stick it is best to return the cylinders to the suppliers rather than to try to force them open.

A gas mask should be available in some other part of the building in case it is necessary to enter the chlorinator room or chlorine store when gas is escaping.

In the last few years there has been a complete change of thought about safety measures. Before about 1980 it was considered that the chances of an accident were negligible and that the amount of gas to be dealt with would be small. It was common practice to give chlorine rooms adequate ventilation at low level, because chlorine is heavier than air, and all doors opened directly to the outside and not to internal rooms.

Most engineers now make provision for a more serious accident, envisaging a large-scale escape of gas, and it is the general consensus that it is advisable

(a) to house chlorine apparatus in gas-tight rooms;
(b) to install sensors that can detect chlorine in the atmosphere;
(c) to provide equipment—with extractor fans, chlorine-neutralizing chemicals and sprays which start automatically when the sensors signal that chlorine gas concentrations in excess of 2 ppm are present in the atmosphere.

Whether the new thinking is better is a matter of opinion. Few engineers seem to be able to visualize the nature of the future catastrophe, with the result that safety solutions vary widely. Whether any of them will work on the night remains to be seen.

In 55 years in the business, the writer has never experienced a major chlorine incident, nor can he recollect hearing of one except for the classic case in Glasgow when a drum fell off the back of a lorry. Even that did not kill anybody.

Testing

The test for free and combined chlorine in the treated water is described in Chapter 14. It is very simple and all attendants should be trained to use it frequently to check the water quality. It is extremely unlikely that any water being distributed for public consumption would contain toxic substances, and in their absence any clear water which shows 0.4 mg/l of free residual chlorine as it leaves the plant can be regarded as safe. Water with more than

1 mg/l of free chlorine, while safe to consume, would be unpalatable and cause complaints. The dosage of chlorine should be of constant concern to every member of the works personnel. Chlorination equipment should be fully duplicated, and no water should leave the works when the equipment is not working.

Chlorine dioxide

If there is even the slightest trace of phenol in the water, chlorine tastes may become very objectionable unless the chlorine is used in the form of chlorine dioxide. This is produced by mixing sodium chlorite with chlorine at a very low pH, which can be attained by using chlorine in a concentration of 500 mg/l and mixing in a ClO_2 generator. Recommended doses of ClO_2 are as follows:

for removal of 0.1 mg/l phenol   0.5 mg/l ClO_2
for removal of 1 mg/l iron   1.2 mg/l ClO_2
for removal of 1 mg/l manganese   2.5 mg/l ClO_2.

Chlorine dioxide does not produce a normal reaction with ortho-tolidine when tested in a comparator. It is usually more than twice as strong as the dose indicated.

11: Waterworks wastes and sludge

Waterworks wastes consist of relatively clear liquids and/or fairly heavy sludges.

LIQUID WASTES

Liquid wastes are normally discharged from base-exchange softening plants which are regenerated with brine. On average about 6% of the total throughput of the softening plant is discharged to waste, but as it is usual to soften to zero only a part of the whole supply, the loss in terms of the daily consumption is modified by the ratio between the amount of water which is softened and the total amount supplied. The main pollutants are the chlorides of calcium and magnesium, in concentrations of about 20 000 mg/l.

Such effluents can generally be discharged to the town sewers if these exist or into a big river or tidal estuary. They are extremely difficult to treat, and where they arise at inland sites in arid countries, care should be taken for disposal to take place under conditions which preclude any possibility of their polluting surface or underground sources.

SLUDGE

Sludge can be defined as a highly concentrated suspension of solids in a liquid. In water treatment plants, sludge can be divided into two main types, depending on whether or not it contains chemicals.

Sludge without chemicals

Sludge without chemicals often arises on plants treating river water and includes sludge discharged from primary settling tanks, 'roughing' filters, microstrainers, and sand washers attached to slow sand filter plants. This type of sludge is relatively inoffensive and can be returned to the river with little treatment or virtually none at all if the river is large. Sludge is of little value and is costly to dewater and store. If the river is large there is little point in trying to save the washwater by recycling, nor will the river suffer unduly if sludge of this type is returned. If the river is small it might pay to save water by concentrating the sludge still further in secondary settling tanks, recycling the supernatant water, and returning the thickened sludge to the stream or treating it on drying beds.

Sludge containing chemicals

All problems of sludge collection, treatment and disposal are increased if chemicals are present. Dewatering is more difficult, recovery of the chemicals is rarely a commercial proposition, and disposal of the partially treated sludge creates a bigger nuisance.

Sludge containing chemicals constitutes the major portion of the effluent from a river abstraction plant. Sludge from the different processes is of different concentrations, that from filter-washing being less concentrated than that from the sedimentation basins. As ejected from either process the water content is at least 99%. The liquid flows easily and does not deposit in the outlet pipework provided that velocities exceed about 1.4 m/s. The total daily volume of sludge is normally 1.5–5.0% of the daily plant throughput. If allowed to accumulate, even in the bottom of a deep settling basin, sludge will eventually putrefy unless chlorine is added. In warm countries where water temperatures exceed 15°C, this becomes a problem and necessitates frequent clearance. Sludge discharge is intermittent and large-diameter drains are required to accommodate maximum rates of flow.

A common way of describing sludge concentration is as per cent W/V, where W is the weight of the suspended solids in grams and V is 100 ml of water. Filter washwater is 0.01–0.1% W/V and basin sludge 0.1–2.0% W/V.

In some plants the filter washwater is returned untreated to the sedimentation basin inlets; in others all the waste water is combined for secondary treatment or disposal as may be necessary.

The settling basin sludge constitutes the main problem. However, where polyelectrolytes are used in the main settlement process, and also where flotation tanks are used, the sludge treatment problem becomes less difficult.

The handling and treatment of sludge is a major problem on any works, and sometimes receives little consideration by plant designers.

In Western countries few river authorities will permit raw or semi-treated sludge to be put back into the river. This has led to the enforced adoption of quite sophisticated sludge treatment techniques. Even so, the accumulation of the dried end-products creates a storage problem and final disposal may have to take the form of land fill. Some lime softening sludges can be recycled, and any sludge in which lime predominates has some small agricultural value, but alum sludge (much the most common) is very little use to anybody.

Silt ejection from settling basins

When silt reaches the bottom of a settling basin it forms a loose flocculant sludge with a high water content. Its volume greatly exceeds its total volume of solids, and the ease with which sludge can be removed from a basin is every bit as important as the efficiency with which the basin causes it to deposit. In fact a basin normally fails because it is precipitating more sludge than is being ejected, and it becomes silt-logged.

If the raw water is relatively clean the sludge can be allowed to accumulate for several weeks and the basins emptied down in turn and cleaned out manually. Where the raw water has suspended solids in excess of 250 mg/l for appreciable periods, daily hydraulic or mechanical extraction should be regarded as essential. In upward-flow tanks with hopper bottoms the sludge can be collected at mid-depth in specially provided cone-shaped pockets and ejected under hydraulic head very easily. However, in cases of high suspended solids concentration, the bleed pipes may have to be increased in diameter (say from 20 mm to 50 mm) and operated continuously. The water losses can then become quite high; losses of up to 15% of throughput have been recorded. This is serious, as it represents losses of dosing chemicals and pumping energy. The overall water losses on a treatment plant should not exceed 3%.

'Flat'-bottomed basins of the ridge and furrow type with perforated collector pipes in the furrows are not now constructed, because although it would appear that sludge could be ejected quite easily, in practice its cohesiveness precludes anything approaching complete removal. When the outlet valves are opened, narrow waterways are pulled through the sludge, and relatively clear water passes to waste, leaving the sludge undisturbed.

The only weakness of the Pulsator, which is otherwise an extremely efficient basin, is the difficulty of moving any sludge lodging at the bottom, because the bottom is obstructed by inlet pipes. Most basins of this type have to be shut down and hand-cleaned. For this reason they are seldom met on softening or iron-removal plants or anywhere where the sludge is voluminous.

Scrapers. The accumulated sludge in a basin can be removed manually (after emptying the basin), or by a mechanical scraper.

Flat-bottomed basins (or those in which the bottom slopes only gently) can be scraped mechanically irrespective of their surface shape. It is thought by some to be economic to install a mechanical scraper when the water habitually carries more than 100 mg/l of suspended solids, and certainly scrapers become essential where the suspended solids average 250 mg/l.

In a round basin, the scraper rotates about a central driving pivot which has horizontal arms to which blades are fixed, set at an angle so as to push the sludge towards a central drain, the whole cleansing operation requiring several rotations of the mechanism to effect complete clearance. The velocity of a scraper of any type should not exceed 0.5 m/min. A square tank can use a circular scraper of the above type, but each corner will remain unswept and will collect a small sloping heap of sludge; however, this cannot exceed the size dictated by the angle of repose of the material. By slightly modifying the apparatus the corners can be swept, either by fixing special sweeps as shown in Fig. 11.1, or by mounting the central bearing on runners and driving the scraper by a motor mounted on a rail which traverses the outside top of the wall, the loose central bearing permitting the lengthening and shortening of the supporting truss as may be necessary.

In a fairly narrow, oblong tank the scraper normally spans the width of the basin and runs lengthways from outlet to inlet end, dragging the sludge towards hoppers quite near to the inlet diffuser wall. To overcome the starting load, this type of scraper needs a bigger prime mover than does a rotating scraper. Where there are many basins lying parallel, a single scraper can be used and transferred from basin to basin by a carriage mounted on transverse rails.

Scrapers of the continuous belt type are commonly found in plate or tube type settlers or other basins where complicated equipment in the upper part of the basin precludes travelling arms. In a plate type settler the cross-wall that deflects the flow through the plates

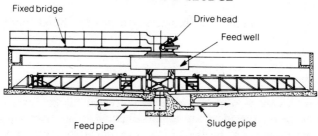

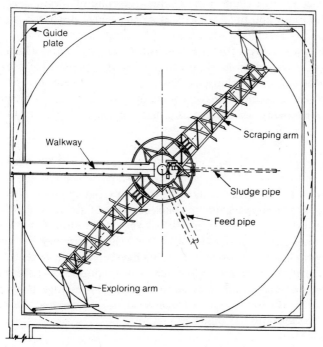

Fig. 11.1. Rotary scraper in a square tank (by courtesy of Dorr Oliver Inc.)

would also foul the belt; to avoid the wall the belt may have to be installed to surface downstream of the plates.

Most scrapers are of proprietary make and are bought and fixed under a separate contract. They are not difficult to make on site but may be covered by patents and for this reason are generally purchased from the makers for local assembly.

There is minimal loss of water when scrapers are used because it is unnecessary to empty the basins or to operate bleed pipes continuously.

Disposal of sludge

Disposal direct to river. Disposal of untreated sludge direct to the river is by far the most satisfactory method if the river authority makes no objection. In developing countries it should be possible, particularly where the river is large, recycling of the washwater is unimportant and nuisance caused by the sludge discharge is small.

Discharge to sewers. In more sophisticated communities which have sewers, discharging the sludge to the sewers is permissible provided that the sludge concentration is not too high. Where the system is practised it seems to have caused less trouble than anticipated, although many sewerage authorities view it with understandable lack of enthusiasm.

Lagooning is a traditional method of sludge treatment but is far from trouble-free. As the average depth of water on the lagoons is only about 1 m, very large areas of land are covered and even after long periods of storage maximum solids contents of about 10% W/V are rarely exceeded. The end product is very sticky stuff, but it can be dug and carted away. The volume of sludge retained in lagoons should normally represent about 200 days' effluent from the settling basins. If this averages 2% of daily plant throughput, the sludge stored in lagoons should be about 4 times the average daily plant capacity. In practice such an area is rarely provided and the results are consequently less than satisfactory.

Concentration and drying. Use of concentrating tanks and drying beds is fairly common. The sludge is discharged to secondary 'fill and draw' settling basins. Two of these are invariably provided: on any given day, one is filling and the other, after standing quiescent for about 6 h, is being pumped clear to be available for filling on the following day. Pumping is by two sets of pumps: one set removes clear supernatant water from the surface by means of a floating-arm suction pipe and delivers it to the sedimentation basin inlets; the other set pumps sludge from the bottom of a sludge hopper at the deepest end of the tank, and this thickened sludge (about 4% W/V)

is delivered to drying beds. The sludge settling tanks are relatively small, each one having to accommodate only one day's supply of washwater, which is equivalent to about 30 min plant throughput. The tank bottoms slope steeply (at about 30° to the horizontal) towards the hopper. In a 10 000 m³/day works, the capacity of each tank would be about 200 m³, and in bigger works would be scaled up proportionally.

The concentrated sludge is discharged to the drying beds, which are somewhat similar to lagoons except that they have sand and gravel bottoms with underdrains. This improved form of construction, together with the fact that the concentrated sludge represents only about 0.25% of the daily output of the works by volume, reduces the problem to more manageable proportions, but even so sludge drying is difficult and the end product needs carting to a tipping place.

In the UK a typical set of drying beds might have an average area of 60 m² per 1000 m³/day, but this would vary upwards or downwards according to local temperatures and rainfall. In a very wet climate the drying beds should be covered.

The sludge has virtually no manurial value but can be used for land fill using the layering methods employed in garbage disposal by municipalities.

Other methods. More sophisticated methods for sludge treatment are vacuum filtration, centrifuging, freezing and filter-pressing, but they are expensive and should be avoided if at all possible in developing countries.

Sludges from precipitation softening. The sludge from lime softeners can be treated and the lime recovered provided that it does not contain too much silt or coagulant. Lime-softener sludge has certain agricultural merit. After removal from the drying beds, it may be acceptable to the local farmers. It is essential to get rid of it or reuse it otherwise it will rapidly occupy the entire plant site.

12: Additional treatment processes

This chapter is written with special reference to the needs of developing countries where the emphasis lies on the supply of water for domestic purposes rather than the special needs of industry. Even so, it is often necessary to remove excessive amounts of iron, manganese or hardness or to correct corrosive tendencies. (Corrosion is discussed in Chapter 13.) Most communities also have private or public swimming pools which are fed from the town mains and need to be kept in reasonable condition.

Of increasing concern to the health authorities is the attention now paid to fluoridation. It is accepted that the teeth of children may be affected by waters low in fluorine and apparatus for adding this element is becoming extremely common in treatment plants in Western countries.

More specialized processes for the reduction or removal of excessive sulphates and chlorides would necessitate the purchase of sophisticated equipment from specialist manufacturers, and are not discussed in detail here. Sulphates give no trouble in concentrations below 250 mg/l and can normally be tolerated if not in excess of 400 mg/l. Chlorides may impart tastes if above 600 mg/l. In most natural waters the concentrations are lower than those mentioned above, but if the only source available did require that they should be reduced, a reverse osmosis plant might be required. Reverse osmosis is a process on which most water engineers would seek advice from specialist manufacturers.

Activated carbon will remove tastes and odours but if dosed as a powder must itself be removed by filtration.

High pressure boilers in power stations and industrial plants require feedwater which is virtually free of dissolved solids. Water of this quality is rarely found in nature and would require to be demineralized. It would be unreasonable to expect treatment to such a high standard to be carried out at source. Such specialized treatment would be applied specifically to the boiler feed water by a demineralizing plant built at the works for which it was required, and its operation would lie outside the normal duties of the engineer in charge of the local waterworks.

Few towns of any consequence in hot countries fail to have an ice-making factory, and as the main raw material of such a factory is probably mains water, certain queries are apt to arise. Ice should be transparent, but all too often it is opaque and although this in no way detracts from its quality it lacks customer appeal. Cloudiness is due to high bicarbonate hardness which may also leave a scum when the ice melts, and can make it brittle at low temperature. Iron causes the ice to be discoloured, and permanent hardness causes whiteness in the core of the frozen block. Sodium salts have much the same effect as non-carbonate hardness. These are all to be expected and cannot be avoided unless the mains water is naturally low in the relevant dissolved solids, or has been treated for their removal.

WATER SOFTENING

Hardness is defined as the effect of any particular water on soap: the less soap it takes to produce a lather, the softer is the water. The salts of calcium and magnesium react with the soap, and until they have been neutralized a lather cannot form. Thus the greater the concentration of calcium and magnesium present in solution, the more soap is needed before a lather will form and the harder the water is said to be.

Hardness is measured by adding measured quantities of soap solution to measured samples of the water under test, and increasing the amount of soap solution until a lather appears on shaking.

There are two kinds of hardness, temporary and permanent, referred to as carbonate and non-carbonate hardness respectively. The temporary hardness precipitates on boiling and is the hardness due to the bicarbonates of calcium and magnesium. It causes the white deposit seen in kettles. The permanent hardness deposits more slowly and is due to the sulphates, nitrates and chlorides of calcium and magnesium.

Hardness is often stated in degrees or parts per million expressed as $CaCO_3$. The old Clark's degree (British) is 1 part in 70 000 (i.e., 1 grain ($\equiv$ 1/7000 lb) per gallon). The French degree is 1 part in 100 000 and may still be used in Europe. The German degree (little used) is approximately 1 part in 56 000. All are being superseded by the practice of expressing results in milligrams per litre. Most degrees are stated in terms of $CaCO_3$ but the German degree is equivalent to 1 part in 100 000 of calcium oxide ($=$ 1 part in 56 000 of $CaCO_3$).

No strict limits to the acceptable degree of hardness have been laid down but 200 mg/l would be regarded as a reasonable upper limit by most people. The widespread use of detergents in recent years has diminished the significance of hardness to the domestic consumer. A statistical but as yet unexplained relationship between soft water and cardio-vascular diseases has been established which has further reduced the trend towards softening.

There are two basic methods of softening water: by precipitation and by base exchange.

Precipitation methods

The hardness-producing salts are in solution. They include calcium bicarbonate, calcium sulphate, magnesium bicarbonate and magnesium sulphate. There may also be the nitrates and chlorides of calcium and magnesium. By adding lime, and soda ash if necessary, non-soluble carbonates can be formed and these can be settled out and filtered in the same way as the silt in a turbid river water. It is common practice to add alum also as a coagulant, its purpose being to bring down the finer particles. Other coagulants can be used with equal success and sodium aluminate is frequently used at high pH values to precipitate magnesium hydroxide.

As carbonate hardness is the most common form of hardness it may only be necessary to add saturated lime water. Where sulphates also have to be moved, both lime and soda may be needed. These can be dosed simultaneously. The reactions are given in Chapter 4 and the plant design parameters in Tables 6.4 and 8.1.

All precipitation methods of softening imply a need for a reasonable knowledge of water chemistry, and the precipitated carbonates also cause a sludge disposal problem. For these reasons the methods may not be particularly suitable for adoption in developing countries, although a number of successful plants are being operated under fairly adverse conditions. In this connection the Spiractor is easy to operate and produces sludge which is hard and inoffensive. It loses effectiveness, however, where the magnesium content exceeds 20–25 mg/l because magnesium salts do not react quickly enough to suit this type of plant.

Base exchange

The base-exchange method of softening is probably the most suitable for use in rural locations because it is simple to operate and is normally purchased as a 'package deal' from a specialist firm.

The hardness-producing salts of calcium and magnesium are replaced by salts of sodium, which do not cause hardness. The following reactions occur:

$$Ca(HCO_3)_2 + Na_2(zeolite) \rightarrow Ca(zeolite) + 2NaHCO_3$$
$$CaSO_4 + Na_2(zeolite) \rightarrow Ca(zeolite) + Na_2SO_4$$
$$Mg(HCO_3)_2 + Na_2(zeolite) \rightarrow Mg(zeolite) + 2NaHCO_3$$
$$MgSO_4 + Na_2(zeolite) \rightarrow Mg(zeolite) + Na_2SO_4$$

After the sodium has been exhausted by base exchange it can be regenerated by washing with a brine solution:

$$Ca(zeolite) + 2NaCl \rightarrow CaCl_2 + Na_2(zeolite)$$
$$Mg(zeolite) + 2NaCl \rightarrow MgCl_2 + Na_2(zeolite)$$

During this process the calcium and magnesium chlorides are discharged to waste.

'Softening to zero' is effected. Zero-softened water is corrosive because of an unfavourable pH/alkalinity ratio, and so it is standard practice to soften only a portion of the raw water, and by mixing this with untreated water to achieve a raw/treated blend which, by retaining part of the hardness-forming salts, will be non-corrosive, reasonably soft and cheaper than a fully treated supply.

This method of water softening may prove to be a little more expensive than lime softening, and does nothing to reduce the total dissolved solids. It works well only on clear water with a low iron content (< 0.5 mg/l). Chlorination should not precede softening. Base exchange will not reduce high concentrations of sulphates or chlorides, and all treated water should be checked to see that these remain at acceptable levels.

However, the method is widely used because of its simplicity and the fact that it produces no solid sludge.

Design. Although most plants are bought from manufacturers on a package deal basis, they are, in fact, easy to design given the necessary technical information about the base-exchange material, which is readily forthcoming from the manufacturers of the resins. As for many other items of a waterworks treatment plant, considerable money can be saved if the plant is locally built using base-exchange material purchased in bulk.

Advancement in softening practice is still proceeding and it is essential to obtain up-to-date data from manufacturers of softening media regarding their product's capacity to remove hardness, and its needs during the regeneration process.

However, in recent years water softening has tended to fall out of fashion. Few new plants are now being built, and one's contact with the science is most likely to take the form of having to operate an existing plant constructed along the lines commonly in use prior to 1980.

The softening units look very similar to pressure filters, but instead of sand a bed of zeolite about 80 cm thick rests on about 45 cm of graded gravel. Downward rates of flow are commonly about 15 m/h and regeneration rates about 20 m/h.

The salt is mixed as a 25% saturated brine solution but is let down to about 5% before being used in the regeneration process. Regeneration takes about 30 min. Before it is returned to operation the bed should be rinsed with raw water for about 20 min.

The plant is generally placed on a bypass to the main raw water supply so that blending to any desired extent can be effected.

The raw water should be clear and virtually free of iron, manganese and chlorine. Base-exchange softeners should serve no purpose other than softening. If silt or iron is present they will be damaged.

They are so easy to operate that, where the mains supply is hard, small units working on the same principle are widely used in private houses and operated by housewives.

Salt storage. Several tonnes of salt have to be stored, either in granular form in bags or in tanks as a 25% brine. This latter solution is corrosive and concrete tanks would need to be protected by some suitable facing material. Injection to the washwater mains should be by hydraulic injector designed to reduce the applied brine to a 5% solution by adding the water used to activate the injectors. Alternatively, dilution can take place in a tank and the brine can be pumped through the softening shells.

Waste disposal. The liquid waste is rich in magnesium and calcium chlorides and should not be discharged into any stream which does not provide at least 50:1 dilution. Great care in disposal is required to ensure that there is no pollution of surface or underground sources, as the salt concentrations may be as high as 20 000 mg/l in the liquid leaving the plant.

REMOVAL OF IRON AND MANGANESE

Iron and manganese cause discoloration of water, stain laundry, impart tastes and encourage the growth of certain bacteria. On the subject of their removal, Cox[1] makes essential reading.

Iron

If the iron content of water exceeds 0.3 mg/l it should be regarded as objectionable and efforts should be made to remove it. It exists in solution in the ferrous state, often as ferrous carbonate, and can only do so in the absence of oxygen and generally where the pH is below 6.5. Because river waters are well oxygenated it is rarely found in solution in surface streams, but is not uncommonly found in solution in deep boreholes or the lower zones of impounding reservoirs. There have been examples of treatment plants designed to treat certain iron-free stream waters which were adversely affected after the construction of an impounding reservoir on the stream because the iron in solution increased.

Treatment for the removal of iron consists of introducing oxygen and raising the pH as might prove necessary. This changes the iron from the soluble ferrous state to the insoluble ferric state $(Fe(OH)_3)$. The iron precipitates and can be settled out and filtered in the usual way. Settling basins can be omitted if the iron content does not exceed 2 mg/l as the filters can operate unaided at such concentrations. Iron is not normally difficult to remove except in the rather rare instances where it has combined with organic matter.

Oxygen is introduced either by aeration or by injection of chlorine, the former being preferable.

Manganese

Manganese often occurs in water for the same reasons as iron and therefore the two may be present simultaneously. Like iron it stains laundry, the stains being black instead of rust coloured. The permissible limit is 0.3 mg/l but if iron too is present the total of the two should not exceed this amount. Iron is the easier of the two to remove, and so if the total concentration of iron and manganese is too high, but the manganese concentration is less than 0.3 mg/l, it will probably be preferable to remove the iron. The difficulty of removal of manganese is due to its oxidizing less easily than iron. Where manganese has originated in the lower zone of an impounding reservoir it may, unlike iron, spread into the upper zone.

For removal of manganese, the pH should be higher than for removal of iron. Oxidation is required, which can be by aeration or chlorination, but the best oxidizing agent for removing manganese is potassium permanganate, which will precipitate the manganese when the pH is above 6.5.

The oxidation may have to be followed by treatment on a catalytic filter, which consists of filter media containing iron and manganese oxides. Alternatively, zeolite (manganese zeolite treated with manganese chloride) can remove manganese if regenerated with potassium permanganate. Often a normal sand filter will act as a catalytic filter after the bed has become suitably seeded with iron and manganese.

If manganese is a major problem it is advisable to set up a pilot plant and experiment a little along the lines indicated above, as it is not easy to predict the correct process in advance.

CONTROL OF FLUORIDES

Fluorides are sometimes found in natural waters. Their importance relates to their effect on dental decay. In concentrations of about 1 mg/l they are wholly beneficial, and statistics indicate that they reduce decay of teeth, particularly among children. In concentrations above 3–5 mg/l, however, mottling of the teeth occurs and reduction of the fluorides becomes desirable.

Fluorine in excess of 1.5 mg/l (which is regarded as the upper safe limit) can be removed by passing the water through tricalcium phosphate, certain ion-exchange compounds or activated alumina. Although effective, the processes tend to be expensive and the materials difficult to regenerate. The most satisfactory method of coping with water with high fluorides may be to abandon the source and find water elsewhere.

Addition of fluorine is not commonly practised in developing countries, as dental decay is not a widespread problem. If special circumstances dictate that addition is necessary, it is generally accepted that fluoride is best added in the form of sodium fluoride (NaF) on small works, sodium silicofluoride (Na_2SiF_6) on bigger works (say over 5000 m^3/day) and hydrofluosilicic acid (H_2SiF_6) where control is advanced and overall cheapness is desired. The maximum dose should total about 1 mg/l as fluorine. The use of dosers of proprietary manufacture is to be recommended. The powders are often mixed with water in tanks and added by solution dosers, but the dosing plant may differ in each case.

It should not be overlooked that fluorine in excessive doses is dangerous and if a plant is installed it needs to be carefully operated.

SWIMMING POOL TREATMENT

Public and private swimming pools are becoming increasingly common and are often operated by people with little knowledge of waterworks chemistry and procedures. Virtually all such pools have to be filled, and refilled occasionally, from the local public supply mains. In a hot climate the quality of the water rapidly deteriorates but it is neither economic nor practicable to provide for continuous replacement of the water from the public mains, and if the bath is to remain usable the water has to be constantly recirculated and treated.

Treatment consists of filtering the water and giving it a high dose of chlorine. The water is pumped out of the deep end and returned to the shallow end. The rate of pumping should be such as to ensure that the entire contents of the bath are recirculated five times a day if the bath is in constant use.

As the water is reasonably clear, high filter rates (12 m/h) can be maintained. Pressure filters are commonly used. A strainer is fitted before the filters, with sulphate of alumina and soda ash often being used as coagulant and for pH correction respectively. The main essential is a high chlorine dose and a pH about 7.4. The problem in swimming pools is that ammonia tends to build up and only a heavy chlorine dose will destroy this. The residual chlorine content in the pool itself should be at least 1 mg/l. This sometimes annoys the bathers but at least it is safe.

Private pools are normally run rather badly. The filters, if fitted at all, are normally too small. They often work on the principle of diatomaceous earth depositing on porous candles, canvas, or wire strainers. The water in the pool tends to go green with algal growth and attempts are sometimes made to control this with doses of hypochlorite and copper sulphate, usually with only partial success. Frequent emptying, cleaning and refilling is normally found to be necessary.

REMOVAL OF DETERGENTS

Although, strictly speaking, soap could be included in the category, the term detergent is commonly applied to the wide range of powders and liquids now used to replace soap for household uses. One of the merits of detergents over soap is that they do not react with hard water to produce calcium soap-based scum, and thus the nuisance of hard water is mitigated.

Some years ago the increasing use of detergents affected the operation of sewage works and the effluents caused foaming in the rivers to which they were admitted. As a result, a concentration of anionic detergents in excess of 0.2 mg/l is considered undesirable, and 1 mg/l is the maximum permissible. Detergents are rather difficult to remove, although filtration has some effect.

The problem has been reduced by alterations to the chemical composition of detergents, which have been made by their makers. Provided that concentrations in the raw water are limited to the above-stated values, detergents should not inconvenience a modern treatment works.

REMOVAL OF INSECTICIDES, FUNGICIDES AND WEED-KILLERS

The widespread use of pesticides which are sprayed on to vegetation and into the atmosphere has inevitably led to traces of poisonous substances in watercourses. In the UK alone there are 1700 commercial products on the market. In Malaysia the habitual use of arsenical compounds in plantations has resulted in many surface sources showing traces of arsenic, but usually below 0.05 mg/l, which is considered tolerable. Experiments conducted in Switzerland[2] show that only activated carbon will positively reduce pesticide concentrations, and even so certain types of activated carbon are more effective than others.

Enforced control to minimize the concentration of the various poisonous elements in the raw water appears to be the most effective remedy.

LEAD

Lead is seldom present in natural water but soft waters are often plumbo-solvent. As it was common practice until recently to use lead service pipes, there are still many areas where a combination of soft water and old property results in unacceptably high levels of lead in the drinking water. To correct this tendency the water supplied should be treated at source; colour should be removed and lime should be added.

NITRATES

Nitrates are not in themselves harmful to adults, but water given to babies below the age of 12 months should not have a nitrate concentration above about 45 mg/l. (Figures given for the limit may vary between 10 mg/l and 80 mg/l.) Water high in nitrates is apt to cause infantile methaemoglobinaemia (blue babies). Nitrates are very difficult to remove and sources containing them in excess should be avoided. In areas where their presence is widespread the only practical solution would appear to be to ensure protection for infants (by breast-feeding or use of bottled water) and to try to alter the agricultural practices from which nitrates arise.

APPROPRIATE TECHNOLOGY

In recent years much use has been made of a wide range of simple practices collectively known as 'appropriate technology'. These are cheap and often ingenious methods of upgrading water supply facilities in backward, tribal areas where lack of money and expertise precludes building to normal standards. At the lowest level this might amount to little more than cleaning up the point where buckets are dipped in a local stream, and ensuring that drinking water is taken from a point above the place where garments are washed. There are hundreds of cases where improvements have been achieved above and beyond the previous lamentable levels and these deserve the highest credit and encouragement. However, the limitations of 'appropriate technology' have never been as widely publicized as its undoubted successes. It is extremely easy to fritter away the meagre resources on inadequate measures which rapidly fall into disuse.

13: Prevention of corrosion

The corrosiveness of water causes immense damage, particularly to metallic surfaces. The causes of corrosion are many and they are still only partially understood. The subject is a wide one, covering cause and protection as well as treatment of the water to reduce its corrosiveness. Unless iron and manganese are known to be present the appearance of discoloured water at the house taps is a sign that corrosion is taking place. Corrosion is an electrolytic action in which the dissolved oxygen plays an important role.

Corrosive tendencies in water result from the presence of dissolved oxygen, and the severity of the attack depends on the pH, the temperature, and the concentration of certain mineral salts. A high hydrogen ion concentration (low pH) is conducive to a high rate of corrosion. The commonest cause of a low pH is a high concentration of dissolved carbon dioxide gas, but as the effect of CO_2 is counteracted by alkaline salts the pH depends on the ratio between the two.

To reduce the corrosiveness of water, any of the following may be effective:

(a) driving off the dissolved oxygen in vacuum chambers;
(b) deactivating the dissolved oxygen by passing the water through iron chips;
(c) reducing the dissolved free CO_2 by aeration;
(d) adding lime to raise the pH;
(e) adding a suitable corrosion-inhibitor.

In practice the methods commonly used on waterworks are (c) and (d). From Figs 13.1 and 13.2, approximate figures can be taken for free CO_2 and equilibrium pH (pH$_s$) corresponding to equilibrium carbonate alkalinity. When the pH of the water as measured is less than pH$_s$, the Langelier index $I (= \text{pH} - \text{pH}_s)$ is negative and the water is undersaturated with $CaCO_3$ and will be corrosive. If pH$_s$ is

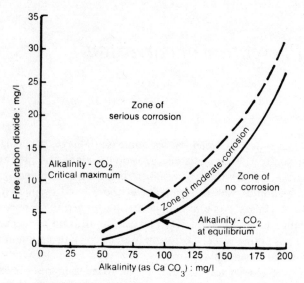

Fig. 13.1. Relationship between alkalinity, CO_2 content and the corrosiveness of water (from Cox[1])

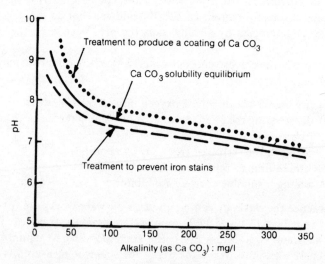

Fig. 13.2. Relationship between alkalinity, pH, calcium carbonate stability and iron staining (from Cox[1])

less than the actual pH then I is positive and carbonate scale may be deposited.

Hydrated lime combines with the CO_2 in solution to form calcium bicarbonate:

$$Ca(OH)_2 + 2CO_2 = Ca(HCO_3)_2$$

When all the CO_2 is neutralized the further addition of lime causes a reaction with the calcium bicarbonate to form insoluble calcium carbonate.

The usual practice is to add enough lime to raise the pH and raise the alkalinity so that the water reaches an equilibrium state where it will neither deposit nor dissolve carbonate scale. The lime should always be added after filtration.

Certain waters are notoriously prone to be corrosive and should be tested immediately. These include soft moorland peat-stained waters, many deep well waters containing CO_2 and iron in solution, water from distillation plants, and water softened by the base-exchange process.

DEZINCIFICATION

Brass is widely used in the manufacture of waterworks fittings. It is an alloy of copper which contains zinc. Zinc is widely used as a protection against corrosion, but can itself be attacked by water containing free CO_2, or having high pH (> 8.2) and a high ratio of chloride to carbonate hardness. In the latter case, for a low chloride content the problem occurs when the ratio of chloride to carbonate hardness is > 1. For chloride > 20 mg/l the problem can occur when the ratio is much lower. As the chloride content is difficult to reduce, the remedy lies in increasing the carbonate hardness. It has been noted, however, that hot-pressed brass is much more susceptible to dezincification than cast brass and an easier remedy might be found in avoiding the use of the former material. Dezincification should always be suspected where a heavy white corrosion product can be seen on brassware in contact with water.

DISSIMILAR METALS

Corrosion is caused by electrolytic action, which is intensified by the presence of dissimilar metals in the same system of water pipes.

Copper is a particularly active metal, especially when in close proximity to zinc or aluminium, and it should be used with care. However, complete installations of copper tubing give little trouble because they have no other metal with which to react. In a floating water treatment plant built entirely of aluminium for service in Iraq some years ago, the danger of using copper was felt to be so great that special taps and valves of aluminium were made for controlling the various items of equipment. In water mains, severe pitting of the iron pipes often occurs in the vicinity of lead joints.

Corrosion problems have been greatly reduced by the increasing use of asbestos-cement, plastic pipes and fittings and rubber joints.

SULPHATE-REDUCING BACTERIA

Some of the worst cases of corrosion occur when sulphate-reducing bacteria are present. These are apt to occur anywhere in the world. They are particularly active in their effect on iron and steel but do not attack these metals directly. They reduce sulphates to form hydrogen sulphide which combines with iron to form graphite. In water their action results in the formation of sulphuric acid. Attack on the metal pipes may be either internal or external according to whether the bacteria are in the water or the soil. Because graphite is smooth and black, the pitting is often difficult to see but it is soft and if suspected can be dug out with a penknife. The corrosion pits have a characteristic circular shape, any irregular areas of attack mainly occurring only because a large number of circles have merged together. The virulence of the attack is due to the continuous nature of the process, the bacteria feeding on by-products of the corrosion products that their activities create.

Sulphate-reducing bacteria are also believed to play an active part in the formation of tuberculation within pipes made of ferrous metal. This phenomenon is serious because the hard-surfaced, hollow lumps which form within a pipeline reduce its carrying capacity often to the point of complete blockage. The nodules themselves provide ideal conditions for the bacteria to flourish.

These bacteria are anaerobic. They cannot live in the presence of oxygen and where they are found in deep well water they can be killed by aeration, which for such water is often needed anyway to remove CO_2 or iron. They are difficult to locate because anaerobic bacteria are destroyed by the ordinary method of taking samples.

Samples should be taken in bottles which have been sterilized and filled with an inert gas, or by special techniques from below water.

Where present in heavy, wet clay soil, they play havoc with external surfaces of iron and steel pipes. To a large extent the problem is one of drainage. Wet areas should be avoided or drained to let the air in, but where this is impracticable special coatings or cathodic protection may be necessary. Their presence should be suspected if a layer of wet clay peeled carefully off an iron pipe exhibits a shiny black surface, almost as if a piece of the coating had come off with it. If, within a few hours, this turns bright orange in colour, sulphate-reducers should be suspected and one should start prodding the pipe surface with a penknife: location of the circular pits of graphite confirms their presence as a matter of certainty.

IRON BACTERIA

Iron bacteria, which flourish in some waters, are aerobic. They can absorb oxygen and oxidize any iron they might encounter, whether in solution in the water or from the pipe itself. If the iron content of the water is high, slimes are formed which affect the potability of the water.

14: Simple laboratory procedures and sampling

TESTING

Tests for colour and turbidity are generally made with special instruments and the makers' instructions should be followed. Turbidity can be measured with a turbidity rod (Table 2.3) but the results thus obtained are approximate. Turbidities over 5 JTUs and colours over 15 mg/l on the platinum–cobalt scale are distinctly noticeable and may lead to complaints from consumers.

Odours are classified in strength from 0 to 5. The figures denote the number of dilutions of pure water to be added before the odour disappears. From 0 to 2 no complaints should arise, 3 and 4 are increasingly objectionable and 5 would indicate that the water is unacceptable.

Hardness can be measured by standard soap solutions, the amount required to give a lather indicating the degree of hardness. Permanent (non-carbonate) hardness is determined in the same way after the water has been boiled to remove the temporary (carbonate) hardness.

Where a qualified chemist is not available, a water testing kit of the Hach type can be used to determine a wide range of dissolved solids, and also pH and residual chlorine.

Jar test

An important piece of apparatus in any treatment plant laboratory is a jar tester. This consists of a motor-driven horizontal spindle driving about four vertical paddles (Fig. 14.1). Each paddle rotates in a glass beaker of about 1.5 l capacity, containing 1 l of the water under test. Stock 1 % solutions of coagulant (often alum and lime) are added in graded doses to each of the beakers. Stirring is initially brisk, then more gentle. The time to form, heaviness, and other characteristics of the floc in each beaker is noted, and after stirring

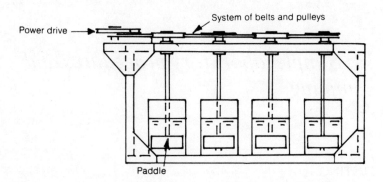

Fig. 14.1. Laboratory stirring device

ceases the settling time is observed. The dosage giving optimum results can then be tried in the plant.

The solutions of the coagulants under test are added by pipette, 1 ml of a 0.1 % solution added to 1 l of water giving a dose of 1 mg/l. The water should be tested before and after the experiment for colour, turbidity and pH, and passed through filter paper to demonstrate the clarity of the end product.

Chlorine test

The test for residual chlorine is important, widely practised and extremely simple to perform. A pocket-sized chloroscope is used in which there are two tubes, each containing a measured quantity of the water under test, which can be compared by eye for colour. One is coloured by the addition of a chlorine-sensitive reagent, the other by a range of standard glass slides. The test consists only of matching the colour of the tube to which the reagent has been added with that of the nearest standard, and the chlorine concentration can be read off directly. Until the 1960s the reagent used was orthotolidine arsenite solution, a few drops of which produce a yellow colour in the sample. Although still widely used, this material is falling out of favour as orthotolidine is believed to be carcinogenic, and it is now being superseded by DPD tablets (NN–diethyl–p–phenylene diamine sulphate) which indicate the strength of the free chlorine in terms of different shades of red. The different proportions of free and combined chlorine can be ascertained by using different types of DPD tablet as recommended by the suppliers.

Tests for other dissolved solids

Similar tests can be used to ascertain the presence and concentrations of other dissolved solids using different indicators and standard slides. These are supplied for field and laboratory use by various makers (e.g., Hach Chemical Co.).

Testing the pH value

The test for pH is performed in somewhat similar fashion to that for residual chlorine. A small pocket comparator is used, a few drops of the testing solution are introduced into the sample, and the colour which develops is compared with standard slides, generally mounted in a disc.

Obviously the testing liquid and colour slides are different from those used in a chloroscope.

There is a small complication in that most liquids used for the measurement of pH are effective only over a narrow range. Fortunately, at any given works the pH does not normally vary much and one or two different test liquids will cover the whole annual range. At an unfamiliar works, the approximate pH must first be ascertained using a wide-range indicator, and once this is known the appropriate narrow-range indicator can be selected for everyday use.

TAKING SAMPLES

The method of collecting samples for laboratory analysis depends on whether chemical or bacteriological analysis is required. For chemical analysis a large sample is required. The bottle, commonly called a Winchester quart, is about 2 l. It should be filled and washed out with the water being tested. It should then be emptied, refilled and securely stoppered. The label should record the time, date and origin of the sample and the name of the person who took it.

Bacteriological samples have to be taken with much more care. The sampling tap should be run, and then to sterilize it a piece of cotton wool should be impregnated with methylated spirit, ignited and held underneath the tap on a piece of wire. The sterilized bottle, which is much smaller than a bottle used for sampling for chemical analysis, should be filled once only and the stopper inserted, great care being taken to ensure that the fingers do not touch the part of the stopper in contact with the sample. (An appreciable percentage of good samples are accidentally contaminated by faulty technique.) The bottle should be packed in ice for its journey to the laboratory.

If a sample tested for coliform bacteria shows positive results where there is good reason to suppose that the water is pure, a check sample should immediately be taken. It often happens that the sample has been carelessly taken and the fault lies in the sampling technique rather than the water quality. If, on checking, the second sample is also positive, swift appropriate action is necessary, with investigations spreading outwards from the affected tap until the point of pollution is located. If the trouble appears to be confined to a single house, the nearest local storage tank should be suspected. Dead birds or rodents in an unprotected tank are a common source of trouble. If bad samples are more widespread, an increase in the dose of chlorine at the works should be made as a precaution while the investigation is proceeding.

Appendix 1: Typical installations

The following pages give details of water quality and treatment processes for plants in the UK and abroad. Post-chlorination is to be assumed in all cases.

		Turbidity, silica scale	Colour, cobalt scale or degrees Hazen	Total dissolved solids, mg/l	pH
River Thames at Oxford (Farmoor) after storage[1,2]	High	11	25	415	9.1
	Low	0.5	0	265	8.8
River Severn at Upton-on-Severn[3]	High	3500	250	538	8.9
	Low	14	6	72	6.8
River Severn at Hampton Loade[4,5]	High from river	840	450‡	—	8.9
	Low from river	4	22‡	—	7.1
	Average after storage	14	55‡	—	7.9
River Derwent at Draycott[6]	High from river	1050	80	552	8.1
	Low from river	5	15	241	7.3
	Average after storage	24	16	430	8.0
River Great Ouse below Buckingham	Typical after storage	5	15	375	8.3
River Great Ouse at Offord[7,8]	High from river	60*	55	500	8.8
	Low from river	2*	5	400	7.3
	After storage	1.5*	5	—	8.3
River Test at Testwood (Southampton) (range of values given)	Normal	{3 30}	<5	{300 340}	{7.9 8.2}
	After heavy rain	{30 100}	{5 10}	—	{6.9 7.9}
Pitsford Reservoir, Northants[9]	Typical	5	15	—	7.9
River Neath (Port Talbot supply)		12†	20	110	7.4

* Paterson.   † APHA.   ‡ Burgess.

Nitrates (N), mg/l	Nitrites (N), mg/l	Dissolved oxygen, % saturation	Free CO_2, mg/l	Chlorides (Cl), mg/l	Alkalinity ($CaCO_3$), mg/l	Total hardness, mg/l	Permanent hardness, mg/l	Temporary hardness, mg/l
						Note: limiting values of permanent and temporary hardness do not necessarily coincide to give limiting value of total hardness		
— —	— —	— —	— —	33 25	199 90	296 178	147 72	190 70
3.5 0	0.3 0	145 40	— —	192 12	220 25	325 60	— —	— —
— — —	— — —	— — —	— — —	62 10 28	155 26 102	224 38 144	— — —	— — —
9.57 1.76 3.60	0.52 0.03 0.10	— — —	— — —	94 20 55	— — —	298 116 235	— — —	— — —
0.29	0.005	—	—	17	160	252	92	160
15 0 0.3	0.3 0 0.1	— 100 110	— — —	80 26 54	— — —	435 285 280	250 85 130	250 155 150
4.4 —	0.006 —	— —	— —	{14 16} —	{215 230} —	{245 260} {165 245}	{30 35} —	{215 230} —
—	—	—	—	19	185	260	75	185
1.1	<0.01	—	—	10	40	70	30	40

Key.
T　Trace

		Iron, mg/l	Manganese, mg/l	Phosphates, mg/l	Oxygen absorbed (4 h at 27°C), mg/l
River Thames at Oxford (Farmoor) after storage[1,2]	High	0	—	1.1	4.5
	Low	0	—	0	0.8
River Severn at Upton-on-Severn[3]	High	—	—	—	5.5
	Low	—	—	—	0.9
River Severn at Hampton Loade[4,5]	High from river	—	—	—	—
	Low from river	—	—	—	—
	Average after storage	—	—	—	—
River Derwent at Draycott[6]	High from river	11.2	1.6	—	15.7*
	Low from river	0.14	0	—	1.6*
	Average after storage	0.18	0.04	—	4.3*
River Great Ouse below Buckingham	Typical after storage	0.04	—	—	2.6*
River Great Ouse at Offord[7,8]	High from river	—	—	—	7.4
	Low from river	—	—	—	2.0
	After storage	0	—	T	2.0
River Test at Testwood (Southampton) (range of values given)	Normal	{0.05 0.2}	<0.02	{<0.005 00.2}	{0.7 1.5}
	After heavy rain	{0.2 2.5}	{0.02 0.25}	—	{1.5 7.0}
Pitsford Reservoir, Northants[9]	Typical	—	—	—	3.3
River Neath (Port Talbot supply)		0.48	0.14	—	1.9

* 3 h at 37°C.

Coli aerogenes	E. coli type	3 days at 20°C	2 days at 37°C	Raw water storage	Coarse screens	Fine screens	Aeration
Bacteria, most probable number per 100 ml		Colonies on nutrient agar per millilitre					
1100 0	460 0	— —	— —	140 days	Yes	—	Overflow weirs
>180 000 700	70 000 170	— —	— —	No	Yes	No	No
210 000 0 143	14 000 0 103	>50 000 0 265	>50 000 0 39	12–15 days	Yes	No	No
{78 200} 800	{67 550} 200	— —	— —	>15 days	Yes	Cup screen, 1600 apertures per in²	Cascade
—	—	—	—	Yes	—	—	—
{6 500} 20	{1 700} 10	{50 000} 580	{15 000} 140	200 days	At intake	—	Splash trays
— —	— —	— —	— —	No	Yes	2 band screens	No
—	—	—	—	Yes	Yes	No	No
—	—	—	—	No	Yes	Yes	No

	Detritus tanks	Pre-chlorination	Microstrainers	
River Thames at Oxford (Farmoor) after storage[1,2]	Yes (13.5 cm/s)	Yes	Yes	N
River Severn at Upton-on-Severn[3]	No	—	—	U flo
River Severn at Hampton Loade[4,5]	No	No	No	N
River Derwent at Draycott[6]	No	Provision made	cf. Fine screens	N
River Great Ouse below Buckingham	—	4.2–5.4 mg/l	—	N
River Great Ouse at Offord[7,8]	No	No	No (provision made)	N
River Test at Testwood (Southampton)	No	Yes	No	N
Pitsford Reservoir, Northants[9]	No	No	No	N
River Neath (Port Talbot supply)	No	No	No	N

Mixing	Flocculation	Type of main settling basin (note: HBUF = hopper-bottomed upward flow)	Rating of main settling basin	Coagulants used	Coagulant aids	Type of filters (note: RG = rapid gravity)
byrinth	In basin	Precipitator (3¼ h horizontal basin on old works)	6 ft/h; may be uprated	Alum	Occasional activated silica (activated carbon)	RG sand (later modified)
ash xers 2 points	No	HBUF	1.2 m/h; with coagulant aids 1.8 m/h	Alum and occasional lime	Activated silica, 1–2.5 mg/l	RG sand
cidental. age 2: sh xing	Incidentally provided by settling basins	Stage 1: Accentrifloc. Stage 2: Pulsator	Accentrifloc 10 ft/h; Pulsator 11.5 ft/h (both can work faster)	Alum	Magnafloc LT 22, 0.008 mg/l	RG sand
ash xers	No	HBUF	7.3 ft/h	Caustic soda for softening; chlorinated copperas	None	RG sand
	—	Accelator	—	Copperas (35–40 mg/l) and chlorine (4.5–5.0 mg/l)	Powdered lime (weighter)	RG sand
ash xer	No	HBUF	5 ft/h normal; 15.3 ft/h softening; 18 ft/h with coagulant aids	Ferric sulphate or alum; lime for normal clarification	Provision made	RG sand and anthracite
cidental	Incidental	Accentrifloc	14 ft/h; 2 basins of 42 ft dia.	Originally ferrous sulphate; later (summer only) alum	Polyelectro-lytes tried —helped settling but caused gelling on filters	RG anthracite and sand
cidental	No	HBUF	8.8 ft/h (softening)	Alum; lime; soda ash	No	RG sand
idental	—	Horizontal	4 h	Aluminium sulphate	No	RG sand

	Rating of filters	Scrapers	Sludge and washwater recovery
River Thames at Oxford (Farmoor) after storage[1,2]	90 gal/ft² per h	—	Yes
River Severn at Upton-on-Severn[3]	91 gal/ft² per h	—	—
River Severn at Hampton Loade[4,5]	200 gal/ft² per h	—	—
River Derwent at Draycott[6]	100 gal/ft² per h	Not required	Yes
River Great Ouse below Buckingham	65 gal/ft² per h	—	Yes
River Great Ouse at Offord[7,8]	150 gal/ft² per h	Not required	Sludge concentrators and lagoons
River Test at Testwood (Southampton)	—	—	—
Pitsford Reservoir, Northants[9]	120 gal/ft² per h	No	Yes
River Neath (Port Talbot supply)	100 gal/ft² per h	No	—

Output	Remarks
—	Problems of algae and taste
—	—
—	Seasonal variation of average water quality after storage, mildly reflecting river quality. Precise control of LT 22 is critical; correct dose gives major saving in alum dose and improved works performance
—	Settling basin performance greatly improved by installation of Gravilectric cones
—	Taste removal: activated carbon filters, 300 gal/ft² per h
50 mgd	—
Phase 1: 5 mgd partially treated and 5 mgd fully treated	Iron carried over at 14 ft/h rise rate, alum floc too light in summer—two new Accentriflocs 61 ft dia. to reduce rise rate to 9 ft/h
—	Calgon used as stabilising agent
—	Alum carried over

Key. C Clear H High L Low T Trace		Turbidity, silica scale	Colour, cobalt scale	Total dissolved solids, mg/l	pH
Sutton Bingham Reservoir	High Low	L L	L L	374 274	8.0 6.6
Tittesworth Reservoir, Staffs[10]	High Low	103 25	75 25	350 110	7.5 6.7
River Wear at Burnhope Reservoir[11]	High Low	L L	200 40	— —	7.6 6.6
River Severn at Purton	High Low	110 33	37 16	501 407	8.0 7.7
River Rother at Hardham	High Low	210 3	50 <5	313 138	9.3 7.2
River Itchen at Portsmouth intake		1.8*	1.0	—	8.3
Sand borehole at Battlesden, Bedfordshire		0	—	281	6.5
Han River at Gooi, Seoul, Korea	High flood Normal	2 000† 4†	14 0	— —	6.9 7.3
Gyobyu, Rangoon	Stored typical	5†	20	—	7.1
River Karaj at Tehran, Iran	Maximum reading Minimum reading	57 000‡ 5†	L L	248 132	8.3 7.5
River Garinono at Sandakan, Sabah	High from river Low from river	400* 110*	— —	H 33	6.5 5.9
Sungei Putat, Malacca, Malaysia	High Low	H C	>70 5	2450 60	6.6 4.0

* APHA units.   † JTUs.   ‡ Suspended solids.

Nitrates (N), mg/l	Nitrites (N), mg/l	Dissolved oxygen, % saturation	Free CO$_2$, mg/l	Chlorides (Cl), mg/l	Alkalinity (CaCO$_3$), mg/l	Total hardness, mg/l	Permanent hardness, mg/l	Temporary hardness, mg/l
						Note: limiting values of permanent and temporary hardness do not necessarily coincide to give limiting value of total hardness		
1.8 / 0	0.2 / 0	— / —	— / —	— / —	260 / 120	280 / 160	70 / 30	255 / 90
2.3 / 0.2	—	—	—	—	68 / 10	135 / 70	—	—
—	—	—	—	—	35 / 10	—	—	—
4.06 / 1.65	0.12 / 0.04	— / —	— / —	40 / 24	216 / 190	302 / 266	94 / 57	—
5.7 / 1.6	0.06 / 0.003	— / —	7 / 0	31 / 23	150 / 62	196 / 112	—	—
4.00	0.019	—	—	17	—	254	35	219
0.08	0	—	42	18	145	192	47	145
— / —	— / —	— / —	1.8 / 1.8	8 / 20	19 / 31	26 / 64	—	—
—	—	—	—	3	42	46	—	—
— / —	— / —	— / —	—	12 / 4	135 / 78	174 / 92	—	—
1.8 / 0.3	T / 0	—	9 / 7	7 / 3	18 / 5	22 / 20	5 / 4	18 / 16
— / —	T / 0	— / —	—	1230 / 6	24 / —	— / —	—	—

		Iron, mg/l	Manganese, mg/l	Phosphates, mg/l	Oxygen absorbed (3 h at 37°C), mg/l
Key. T Trace					
Sutton Bingham Reservoir	High Low	— —	— —	— —	6.8* 1.2*
Tittesworth Reservoir, Staffs[10]	High Low	2.0 0.2	1.0 0.12	— —	— —
River Wear at Burnhope Reservoir[11]	High Low	— —	— —	— —	— —
River Severn at Purton	High Low	0.22 0.1	0.11 0.03	— —	5.9 2.3
River Rother at Hardham	High Low	0.39 0.23	0.42 0.01	— —	— —
River Itchen at Portsmouth intake		0.07	—	—	0.8
Sand borehole at Battlesden, Bedfordshire		6.28	T	—	0.9
Han River at Gooi, Seoul, Korea	High flood Normal	0.8 0.16	0.8 —	— —	9.2 2.8
Gyobyu, Rangoon	Stored typical	0	0.1	—	—
River Karaj at Tehran, Iran	Maximum reading Minimum reading	<0.01 <0.01	0.05 0	— —	— —
River Garinono at Sandakan, Sabah	High from river Low from river	2.2 0.12	0.6 0.04	— —	— —
Sungei Putat, Malacca, Malaysia	High Low	6.0 1.4	— —	— —	9.4 1.2

* 4 h at 27°C.

Suspended solids, mg/l	Bacteria, most probable number per 100 ml		Colonies on nutrient agar per millilitre		Raw water storage	Coarse screens	Fine screens	Aeration
	Coli aerogenes	E. coli type	3 days at 20°C	2 days at 37°C				
— —	— —	— —	— —	— —	Reservoir	—	—	—
— —	— —	— —	— —	— —	Reservoir	No	No	No
— —	— —	— —	— —	— —	Reservoir	Bar screens	No	—
01 30	>18 000 35	— —	— —	— —	2 × 12 million gal	Yes	Band screens, 3–9 ft/min	2 speed, variable depth; 100 lb O_2/h
20 1	— —	— —	— —	— —	No	—	—	—
—	—	—	—	—	No	Yes	Cup screens, 2 of 11 ft 6 in dia., ¼ in mesh	No
0	—	—	—	—	No	No	No	Spray; 100 m²
— —	460 400	— —	— —	— —	No	Yes	No	Some incidental
—	20	—	—	—	Reservoir	Yes	Yes	Yes
000 3	— —	— —	— —	— —	No	Yes	No	No
— —	— —	— —	— —	— —	Reservoir	Yes	—	No
— —	— —	— —	— —	— —	—	—	—	—

	Detritus tanks	Pre-chlorination	Microstrainers	
Sutton Bingham Reservoir	—	Yes (to inhibit algae)	—	—
Tittesworth Reservoir, Staffs[10]	No	No	No	N
River Wear at Burnhope Reservoir[11]	—	—	—	—
River Severn at Purton	—	2.5 mg/l	No	N
River Rother at Hardham	—	—	—	—
River Itchen at Portsmouth intake	No	8 mg/l	No	N
Sand borehole at Battlesden, Bedfordshire	—	No	No	N
Han River at Gooi, Seoul, Korea	Yes	5 mg/l	No	N
Gyobyu, Rangoon	No	No	No	N
River Karaj at Tehran, Iran	—	2 mg/l before Pulsators	No	<
River Garinono at Sandakan, Sabah	No	No	No	N
Sungei Putat, Malacca, Malaysia	—	—	—	—

Mixing	Flocculation	Type of main settling basin (note: HBUF = hopper-bottomed upward flow)	Rating of main settling basin	Coagulants used	Coagulant aids	Type of filters (note: RG = rapid gravity)
	—	HBUF	0.0011 ft/s	Aluminium sulphate, 40 mg/l	Sodium aluminate, 3 mg/l	—
ash ixers	No	HBUF	4 ft/h	Aluminium sulphate	Occasional lime, chlorine and potassium permanganate	RG sand
jected to main	No	None	—	Aluminium sulphate; occasional ground limestone	Activated silica, 1.5–2.0 mg/l	Rapid—pressure
speed rbine, –50 rpm	No	HBUF	1.46–3.32 m/h; 8 basins of 1369 ft²	Aluminium sulphate, 25 mg/l	Activated silica, 5 mg/l max.	RG anthracite
.	—	Phase 1: upward rectangular. Phase 2: Accentrifloc	Phase 1: 0.43 mm/s. Phase 2: 0.41 mm/s	Aluminium sulphate	Activated silica	RG sand
ash ixer	No	HBUF	5 ft/h; 16 basins of 1220 ft²	Aluminium sulphate	Polyelectrolyte	RG sand and anthracite
o	13 000 gal basin	Circular radial flow	90 000 gal	Lime, 53 mg/l	None	RG sand
ash ixer, 5 min	30 min; 16 units with paddles	Horizontal plated	1 h 20 min; 16 basins	Aluminium sulphate	Slaked lime	Greenleaf; sand, anthracite
es	No	Horizontal	6 h	Alum	—	None
es	No	Pulsator	2.44 m/h	Ferric chloride, lime	None	RG sand
es	No	Proposed HBUF	4 ft/h	—	—	RG sand
.	—	—	—	—	—	—

	Rating of filters	Scrapers	Sludge and washwater recovery
Sutton Bingham Reservoir	—	—	Lagoons, 5000 ft², 2 mgd
Tittesworth Reservoir, Staffs[10]	90 gal/ft² per h	—	2 tanks, each taking 5 filter washes, and lagoons
River Wear at Burnhope Reservoir[11]	60 filters, total output 7 mgd	—	—
River Severn at Purton	150 gal/ft² per h; 8 filters of 954 ft²	No	—
River Rother at Hardham	1.1 mm/s	No	Very high losses until Gravilectric cone fitted
River Itchen at Portsmouth intake	200 gal/ft² per h	No	3 upward flow × 900 ft²
Sand borehole at Battlesden, Bedfordshire	81.25 gal/ft² per h	—	2 tanks of 36 000 gal; 6 h
Han River at Gooi, Seoul, Korea	250 m/day	Ring belt	—
Gyobyu, Rangoon	—	No	—
River Karaj at Tehran, Iran	—	No	Lagoons
River Garinono at Sandakan, Sabah	100 gal/ft² per h	—	—
Sungei Putat, Malacca, Malaysia	—	—	—

Output	Remarks
—	—
—	Problems of iron and manganese, dyeworks effluent
—	Soft, peat-stained moorland water. Longer contact for mixing gave improved results. Breakthrough of alum and short filter runs, generally in late summer; cured by activated silica
—	Also softening tanks: 4×127.2 m^2; rise rate $1.46-5.36$ m/h; using ferrous sulphate, activated silica and lime
160 l/s	Sulphates 26–48 mg/l. Polyelectrolytes tried without benefit. Accentrifloc basins could be used for softening with rise rate of 0.83 mm/s
Phase 1: 15 mgd	This river is of the same type as the River Test. However, it is infinitely easier to treat, being less subject to variation in quality due to its having a greater proportion of inflow from underground sources and less surface run-off during flood
0.84 mgd	Water typical of certain borehole waters in sand beds; high iron and CO_2; corrosive
600 000 m^3/d	Below Gooi intake the Han River becomes very polluted
24 mgd	—
8 m^3/s	Occasional excessive silt
—	Impounding proposed, with anticipated reduced turbidity and increased iron. This works was not completed; the plant was finally installed at Tawau Sabah
—	—

		Turbidity, JTU	Colour, cobalt scale or degrees Hazen	Total dissolved solids, mg/l	pH	
Key. C Clear L Low D Detected ND Not detected						
Johore River, Singapore, Malaysia, intake at Kota Tinggi[12]	High reading Low reading	550 200	180 10	— —	6.3 6.2	
River Tigris at Bagdad	High reading Low reading	>40 000* <100*	— —	— —	8.3 7.5	
Mekong River at Pakse, Laos	High reading Low reading	2240 30	240 10	— —	7.6 7.0	
River Seine at Orly, Paris	Typical	25	40	—	7.8	
River Ottawa, Canada, at Ottawa city intakes	Maximum Minimum Mean	28 1.6 3.3	60 32 42	110 43 68	7.9 6.6 7.3	
Muda River, Malaysia, at Penang intake	Maximum Minimum	— —	70 20	350 55	7.8 6.5	5 1
Batang Kali, Ulu Selangor South, Malaysia	Maximum Minimum	L C	30 5	45 25	7.3 6.4	

* Suspended solids.

Free and saline ammonia (N), mg/l	Nitrates (N), mg/l	Nitrites (N), mg/l	Dissolved oxygen, % saturation	Free CO_2, mg/l	Chlorides (Cl), mg/l	Alkalinity ($CaCO_3$), mg/l	Total hardness, mg/l	Permanent hardness, mg/l	Temporary hardness, mg/l
								Note: limiting values of permanent and temporary hardness do not necessarily coincide to give limiting value of total hardness	
03 / 02	0.5 / 0	0 / 0	— / —	— / —	6 / 1	5 / 3	8 / 3	3 / 0	5 / 3
20 / 10	0.40 / 0	0.10 / 0	— / —	— / —	— / —	155 / 50	285 / 115	140 / 50	155 / 50
D / D	D / ND	D / ND	— / —	— / —	— / —	153 / 70	108 / 5	— / —	— / —
0.4	—	—	—	6.5	2	21	250	—	—
16 / 01 / 07	0.48 / <0.01 / 0.17	— / — / —	— / — / —	— / — / —	176 / 2 / 10	46 / 12 / 23	57 / 16 / 33	— / — / —	— / — / —
27 / 02	0.20 / 0	0.001 / 0	— / —	— / —	12 / 4	— / —	— / —	— / —	— / —
36 / 01	— / —	0 / 0	— / —	— / —	— / —	10 / 5	4 / 2	— / —	— / —

Key.
L Low

		Iron, mg/l	Manganese, mg/l	Phosphates, mg/l	Oxygen absorbed (4 h at 27 C), mg/l
Johore River, Singapore, Malaysia, intake at Kota Tinggi[12]	High reading	1.2	—	—	8.0
	Low reading	1.2	—	—	3.4
River Tigris at Bagdad	High reading	—	—	—	—
	Low reading	—	—	—	—
Mekong River at Pakse, Laos	High reading	0.96	0	—	30.9
	Low reading	0.002	0	—	3.2
River Seine at Orly, Paris	Typical	0.5	—	—	—
River Ottawa, Canada, at Ottawa city intakes	Maximum	—	—	—	—
	Minimum	—	—	—	—
	Mean	—	—	—	—
Muda River, Malaysia, at Penang intake	Maximum	1.60	0.1	—	7.7
	Minimum	0.35	0	—	1.0
Batang Kali, Ulu Selangor South, Malaysia	Maximum	0.34	—	—	3.3
	Minimum	0.1	—	—	0.8

Suspended solids, mg/l	Bacteria, most probable number per 100 ml		Standard plate count, colonies per millilitre		Raw water storage	Coarse screens	Fine screens	Aeration
	Coli aerogenes	E. coli type	at 20°C	at 37°C				
635* 125	— —	— —	— —	— —	No	No	Floating boom and gates	No
000 25	— —	— —	— —	— —	No	No	No	No
— —	— —	— —	— —	— —	No	Yes	Strainers on pumps	No
30	—	—	—	—	No	30 mm	Band screens, 1.4 mm opening	—
— — —	7800 2 187	7000 0 40	— — —	>3000 10 1140	No	Yes	—	No
350 55	— —	— —	— —	— —	In canal and forebay	Yes	Yes	No
L L	— —	— —	— —	— —	No	Yes	No	No

* Volatile suspended solids 95 mg/l.

	Detritus tanks	Pre-chlorination	Microstrainers	Preliminary basins
Johore River, Singapore, Malaysia, intake at Kota Tinggi[12]	No	No	No	No
River Tigris at Bagdad	No	No	No	No
Mekong River at Pakse, Laos	No	No	No	No
River Seine at Orly, Paris	No	8 mg/l	No	No
River Ottawa, Canada, at Ottawa city intakes	No	Yes	No	No
Muda River, Malaysia, at Penang intake	No	No	No	No
Batang Kali, Ulu Selangor South, Malaysia	No	No	No	No

Mixing	Flocculation	Type of main settling basin (note: HBUF = hopper-bottomed upward flow)	Rating of main settling basin	Coagulants used	Coagulant aids	Type of filters (note: RG = rapid gravity)
Reversal through 180° in inlet channel	No	HBUF	9.5 ft/h	Alum, lime	Activated silica	RG sand
In channel	No	Horizontal	4 h	Aluminium sulphate	—	RG and pressure sand
Some	20 min; paddle	2 storey horizontal	105 min	Alum	Lime	RG sand
Yes	No	Pulsator	3.1 m/h	—	Activated silica	RG sand
Yes	40 min	Horizontal	3 h 40 min	Alum and lime	Activated silica	Mixed media (anthracite)
Incidental	Sinuous channel	Horizontal	4 h	Alum	Lime	RG
Incidental	No	Horizontal	4 h	Alum	Lime	RG

	Rating of filters	Scrapers	Sludge and washwater recovery
Johore River, Singapore, Malaysia, intake at Kota Tinggi[12]	150 gal/ft^2 per h	No	No; discharge to river
River Tigris at Bagdad	100 gal/ft^2 per h	Square traction	No; discharge to river
Mekong River at Pakse, Laos	4.3 m/h	No	No
River Seine at Orly, Paris	5 m/h	No	—
River Ottawa, Canada, at Ottawa city intakes	5 m/h	No	No
Muda River, Malaysia, at Penang intake	5 m/h	No	Yes
Batang Kali, Ulu Selangor South, Malaysia	5 m/h	No	No

Output	Remarks
Phase 1: 30 mgd	High colloids caused initial treatment problems
Several 3–11.5 mgd units	Occasional need to slow down at peak turbidities; HBUF tanks always in trouble at these times
315 m³/h	—
30 000 m³/d	Water tends to de-gas on filters
60 mgd	Temperature range 0.5–29°C, mean 11°C. Additional treatment: activated carbon, fluoride
—	Tropical temperatures. Water straw-coloured
3.5 mgd	Tropical temperatures

Glossary

Acidity may be due to natural causes such as the presence of free carbon dioxide or humic acid but it can also result from industrial pollution. Any water with a pH value below 7 is termed acidic. Waters with low pH values are often corrosive.

Activated silica is a highly effective coagulant aid which is prepared from sodium silicate 'activated' by various chemicals which include chlorine, sodium bicarbonate, sulphuric and hydrochloric acid, and carbon dioxide. These act as neutralizing agents on the sodium silicate to form a silica sol which in very small doses (2–5 mg/l) will often greatly improve the performance of the primary coagulant.

Aeration is the practice of exposing small droplets of water to air to encourage the absorption of oxygen from the atmosphere or the release of other gases in order to effect beneficial changes in the water.

Algae are primitive organisms which are usually classified as plants. There are hundreds of different types, many of them microscopic, which may become visible by multiplication. When present to excess they cause trouble by blocking filters. Outbreaks vary with the region and the season.

Alkalinity, when used without qualification, means the total alkalinity determined with methyl orange indicator. It is sometimes written as alkalinity (M), and is not the same as alkalinity (P), which has relevance only in lime–soda softening. Alkalinity is caused by the bicarbonates, carbonates and hydroxides of calcium, magnesium, potassium and sodium. High alkalinity is often associated with high total dissolved solids and high pH. When the alkalinity equals the temporary hardness, no permanent hardness is present. If the alkalinity exceeds the total hardness, sodium bicarbonate must be present.

Alum: a commonly used word for aluminium sulphate, probably the most widely used coagulant.

Anthracite filters: a commonly used term for multi-layer filters in which a top layer of anthracite rests on a sand layer below. The sand grains have a greater density than the anthracite, and after the upheaval of washing the sand settles more quickly and the anthracite remains on top.

Backwashing is the standard method of cleaning rapid sand filters whereby water is passed upwards through the filter media to dislodge, by viscous drag, the dirt accumulated during the filtering process. This action is sometimes preceded by air also being blown upwards through the filter media to loosen the dirt.

Base exchange is a method of softening hard waters in which a natural or synthetic material introduces sodium-based salts to replace those of calcium and magnesium, thus softening water passing through it. The material has to be regenerated at frequent intervals by washing with a brine solution.

BOD (*biochemical oxygen demand*) is used as an indicator of the degree of pollution in rivers. The oxygen content of water decreases as the amount of sewage present increases. The BOD is the oxygen absorbed at 20°C over 5 days. Most rivers selected for public supplies have BODs of 4 mg/l or less.

Brownian motion was investigated by Robert Brown, an English botanist of the early 19th century, who observed that microscopic particles in suspension in a gas or a liquid had a continuous, haphazard, zig-zag motion. Later investigators proved that this was due to elastic collisions between molecules, which were strong enough to affect colloidal particles suspended in a liquid. It is one of the reasons why liquids in the same container tend to mix even if not stirred.

Clarifier: see Settling basin

Clark's process is a method of softening hard water by adding lime to react with the free carbonic acid and the carbonic acid combined in the bicarbonates of calcium and magnesium to form insoluble carbonates and hydroxides which precipitate.

Coagulant: a substance used to precipitate suspended solids in water, usually a salt of aluminium or of iron. The precipitate, referred to as a floc, is basically a hydroxide.

Coagulant aid: a substance which, when added to water containing a coagulant, intensifies the settling action and appears to make the floc denser.

Coefficient of fineness: the ratio between the dry silt by weight in milligrams per litre and the turbidity in JTUs, for any given sample of water. This gives an indication of the ease (or otherwise) of settling the solids in that water. For silt-laden water with a coefficient of fineness above unity, the solids will settle more quickly than for water with a coefficient below unity, because the higher figure indicates the presence of a higher concentration of fine sand.

Colloids are suspended matter in a particularly finely divided state. The dividing line between coarsely dispersed material and colloids is where the particles are just visible under an ordinary microscope, having diameters of less than about 1 μm. Colloids carry small negative electric charges which in relation to their weight are large enough for the particles to repel each other and remain in suspension.

Colour. Many waters have a distinct colour even after all turbidity has been removed. This is expressed in terms of the platinum–cobalt scale (Hazen units).

Discrete particle: as commonly used in discussing the theory of settlement, the term applies to a particle which remains separate and of constant size, weight and shape throughout the whole process. This facilitates theoretical reasoning but rarely accords with practical conditions.

Drag coefficient: a value, C, related to the resistance to movement of a sphere through water. It varies with the Reynolds number, which varies inversely with viscosity which in turn varies inversely with temperature. The practical effect is that the speed of settlement increases with the temperature of the water.

Flash mixer: a chamber in which coagulants are stirred into the raw water with considerable violence induced either hydraulically or mechanically.

Floc is the fine cloud of spongey particles that form in water to which a coagulant has been added. The particles are basically hydroxides, commonly of aluminium or iron. They accelerate the settlement of suspended particles by adhering to the particles and neutralizing such negative electric charges as may be present.

Floc volume fraction: the volume of floc per unit volume of suspension, normally in the range 0.0004–0.001.

Flocculation is the practice of gently stirring water in which floc has formed to induce the particles to coalesce and grow. The bigger and denser the floc particles, the quicker is the rate of settlement.

Flotation tank. The attachment of tiny air bubbles induces the suspended matter to rise in a flotation tank rather than to sink in a settling tank, and removal of the solids is effected by skimming from the surface rather than by scraping from the bottom.

Formazin turbidity unit (FTU): see Turbidity

Froude was an English naval architect of the 19th century who founded the science of predicting hydraulic results from geometrically similar models.

Froude's law, $(Fr) = v^2/lg$, where v is velocity and l refers to dimensions, sets forth the condition of similarity of gravity and inertia forces.

Froude's number, (Fr), should be the same in both the model and the object for the paths of flow to be similar.

G values. The stirring of water creates differences of velocity and therefore velocity gradients. The average temporal mean velocity gradient in a shearing fluid, denoted by G, is a factor used in calculating the size of mixing and flocculating chambers.

Hazen, Alan: an American prominent in the field of hydraulics, who in 1904 first presented the argument that the predominant factor in basin performance was the overflow rate Q/A, where A is the surface area of the basin and Q the throughput, and that detention time was not of primary significance. It should follow that depth is not important; for other reasons, however, depth *is* important in settling basins.

Hazen units are a means of expressing the degree of colour in water and are numerically the same as those of the platinum–cobalt scale.

Jackson turbidity unit (JTU): see Turbidity units

Lime–soda plants are softening plants which use lime and soda ash to make the calcium and magnesium salts insoluble so that they form precipitates which can be settled and filtered out of water.

Multi-layer filters are filters in which layers of different materials with different densities are used so that a coarser (but lighter) layer may remain on the surface. A commonly used surface layer is anthracite.

Nitrate is the final stage of oxidation of waste organic matter to amino-acids and then ammonia and indicates the likelihood of pollution of a source of water.

Nitrite. As an intermediate stage in the process of oxidation of waste organic matter, nitrites suggest the possibility of recent pollution.

Overflow rate relates the amount of water passing through a horizontal-flow sedimentation basin to its surface area. If Q is flow in cubic metres a day and A is area in square metres then the overflow rate is Q/A m/day. A commonly used figure for average conditions is 18 m/day but this may be modified by an experienced designer to suit local conditions.

Platinum–cobalt scale: a scale used for measuring colour in water against different concentrations of a platinum–cobalt solution. One unit of colour corresponds to one milligram of platinum per litre.

Poiseuille was a French physician whose studies of the viscous flow of blood through arteries led later researchers to define the coefficient of viscosity and acknowledge his pioneer work when naming the centipoise (a unit of absolute viscosity).

Pressure filter: very similar in design and function to a rapid gravity sand filter except that it is contained in a steel pressure vessel and can operate under pressure if hydraulic conditions in the system require it to do so.

Protected colloids: a term used to describe colloids which have become coated with sewage as a result of being present in a polluted river. They are extremely difficult to precipitate.

Rapid gravity filter: a development of the slow sand filter which works on a somewhat different principle. Its purpose is to remove the finer particles carried over from the settling basins. It works at high rates and can be washed easily and quickly. It is now widely adopted where filtration is necessary.

Raw water: water in its untreated state as it enters the treatment plant.

Reynolds number, (*Re*). A dimensionless parameter providing a criterion for dynamic similarity in fluid flow experiments. $(Re) = vd/\nu$, where v is the average velocity, d is the diameter (or the hydraulic radius $\times$ 4), and ν is the kinematic viscosity.

Reynolds, Osborne, was a Professor of Engineering at Manchester University who became famous for his work in hydraulics and hydrodynamics.

Schmützdecke is a German word meaning 'layer of dirt'. It is a 'skin' which forms on the surface of a slow sand filter and acts as a fine strainer. It is of complex make-up, consisting mainly of many of the impurities the filter is removing from the water.

Sedimentation basin: see Settling basin

Septic: a term commonly applied to deposited sludge or sewage when it becomes deoxygenated and putrifies. In this condition it smells, turns black and gives off gas.

Settling basin, settling tank, sedimentation basin, clarifier: these terms are synonymous and signify the chamber in which settlement occurs. Such chambers mostly belong to two great families, namely horizontal-flow tanks in which the direction of water flow is predominantly horizontal, and vertical-flow tanks in which the water enters at the bottom and overflows from the top. The former can be further subdivided into deep and shallow settlers, the latter into tanks working with sludge accretion and those without. There are many variations of each type. In addition there are a few spiral-flow basins, and a few surviving fill-and-draw tanks (which are filled and emptied after standing for a period).

Silica scale: see Turbidity units

Slow sand filters are the oldest form of sand filter, still fairly widely used. The water is passed very slowly downwards through sand beds of about 75 cm thickness. After about a month in operation the surface has to be skimmed, but this period can vary widely depending on the condition of the raw water. Eventually the filter has to be taken out of service and refilled with washed sand.

Stokes' law expresses the settling velocity finally attained by a solid falling through a liquid (or the rising velocity for a lighter particle). For particles of concern in water treatment it can be stated as

$$v_s = \frac{g}{18} (\rho_1 - \rho) \frac{d^2}{\eta}$$

where v_s is the final settling velocity, cm/s

ρ_1 and ρ are the densities of particle and water respectively, g/cm^3

g is the acceleration due to gravity (981 cm/s^2)

d is the diameter of the particle, cm

η is the dynamic viscosity of water, g/(cm s).

Stokes, Sir George: one time Professor of Mathematics at Cambridge University whose studies included original work on the friction of fluids in motion and the determination of rates of settlement of solids in liquids.

Suspended solids are the solid particles in the water. They can be filtered out, dried and weighed and are then expressed in terms of milligrams per litre.

Tastes and odours are closely related and arise in water for various reasons. Most waters have a slight taste which regular consumers do not notice. Odour is measured in threshold odour numbers (TONs). The number represents the number of times a sample has to be diluted with pure water before the odour can barely be detected. Numbers 0 to 2 are considered acceptable, 3 and 4 are increasingly unpalatable and 5 and upwards are likely to cause complaint and become progressively unacceptable.

Turbidity is caused by suspended solids but is an optical effect caused by dispersion of and interference with light rays. It cannot be directly related to the quantity of solids present because it is also affected by their size, colour and shape. It is measured in turbidity units in various kinds of turbidimeter or by comparing a tube filled with the water on test with tubes containing standard suspensions of formazin or fuller's earth. Some turbidimeters depend on the point at which a candle flame disappears when viewed through a column of water of variable length. The length of the column can be calibrated to give direct readings in the appropriate turbidity units. Other meters use electric cells to measure the intensity of light scattered by the suspended solids.

Turbidity units. Turbidity is expressed in the units applicable to the method of measurement or the instrument used (e.g., JTUs, FTUs, NTUs, APHA units). For all practical waterworks purposes the units are numerically the same except for units expressed in parts per million on the silica scale. These may be the same as the others if a certain type of silica is used to prepare the standard but are apt to vary if other types of silica are used.

Uniformity coefficient: a term often used to define the grading of filter sands. It is the ratio between the aperture openings passing 60% and 10% by weight of the sand sample.

Upward-flow filters are filters in which the flow and the backwashing process proceed in an upward direction. They have the advantage of being able to store a large quantity of sediment in the coarser lower layers and thus require less backwashing. They are extremely uncommon.

van der Waals attractive forces are attractive forces between molecules which may play a part in the removal of minute particles by a rapid gravity sand filter, the sand grains of which are large in relation to the particles they can intercept.

Viscosity is the resistance of a fluid to flow, sometimes called dynamic or absolute viscosity, η. The dynamic viscosity divided by the density of the fluid is called the kinematic viscosity, v.

Wet silt by volume. It is in this form that silt in water is actually encountered. The measurement of dry silt by weight may be informative for the purpose of comparing different waters, but it gives no impression of the bulky and unpleasant amount of wet silt by volume that collects in basins and inlet channels and has to be removed. For example, in the River Tigris (Iraq), 2200 mg/l of dry silt by weight[1] produce 10 000 ppm of wet silt by volume.

W/V = weight/volume ratio: a method of stating the concentration of solids suspended in sludge. Generally the weight is measured in grams and the volume of water in decilitres, and the result is expressed as $x\%$ W/V.

References

Chapter 2

1. FOX C. S. *The Geology of water supply*. Technical Press Ltd, London, 1949, 61.
2. WORLD HEALTH ORGANIZATION. *International standards for drinking water*. WHO, Geneva, 1971, 3rd edn.
3. SUCKLING E. V. *The examination of waters and water supplies*. J. & A. Churchill Ltd, London, 1943, 5th edn, 213.
4. REID E. F. Water treatment in the tropics. *J. Instn Wat. Engrs*, 1956, **10**, Nov.
5. AMERICAN SOCIETY OF CIVIL ENGINEERS *et al. Water treatment plant design*. American Water Works Association Inc., New York, 1969, 27.

Chapter 3

1. SUCKLING E. V. *The examination of waters and water supplies*. J. & A. Churchill Ltd, London, 1943, 5th edn, 604.
2. CROWE P. J. The Farmoor Source works in operation. *J. Instn Wat. Engrs*, 1974, **28**, Mar., No. 2, 123–124.
3. SAXTON K. J. H. Operation of the Grafham Water scheme. *J. Instn Wat. Engrs*, 1970, **24**, Oct., No. 7, 413.
4. HOLLAND G. J. Extensions to Hampton Loade works. *J. Instn Wat. Engrs*, 1974, **28**, May, No. 3, 164.
5. ADAMS R. W. *et al.* The River Derwent scheme of the Nottingham Corporation. *J. Instn Wat. Engrs*, 1973, **27**, Feb, No. 1, 33.
6. COX C. R. *Operation and control of water treatment processes*. World Health Organization, Geneva, 1964, 212.
7. BULL A. W. and DARBY G. M. Sedimentation studies of turbid American rivers. *J. Am. Wat. Wks Ass.*, 1928, **19**, 284.
8. KINCAID R. G. Special design features of water works facilities for highly turbid waters. *Proc. Am. Soc. Civ. Engrs*, 1953, Oct., separate 309.

Chapter 4

1. COX C. R. *Operation and control of water treatment processes*. World Health Organization, Geneva, 1964.

2. WADDINGTON A. H. and COOK W. J. M. Water treatment for public and industrial supplies. *Chemical engineering practice.* Butterworth.
3. TWORT A. C. *et al. Water supply.* Edward Arnold, London, 1974, 2nd edn.
4. BRANSBY-WILLIAMS G. *The purification of water supplies.* Chapman and Hall, London, 1946.

Chapter 5

1. FAIR G. M. *et al. Water and wastewater engineering.* John Wiley and Sons Inc., New York, 1968, **2**, 26–2 to 26–8.
2. CAMP T. R. and STEIN P. C. Velocity gradients and internal work in fluid motion. *J. Boston Soc. Engrs*, 1943, Oct., 219.
3. HUDSON H. *Water clarification processes.* Van Nostrand Reinhold, New York, 1981.

Chapter 6

1. CAMP T. R. Sedimentation and design of settling tanks. *Trans. Am. Soc. Civ. Engrs*, 1946, **111**, 895–958.
2. FAIR G. M. *et al. Water and wastewater engineering.* John Wiley and Sons Inc., New York, 1968, **2**.
3. IMHOFF K. *Disposal of sewage.* Butterworth, London, 1971.
4. COX C. R. *Operation and control of water treatment processes.* World Health Organization, Geneva, 1964, 81.
5. CULP G. L. and CULP R. L. *New concepts in water purification.* Van Nostrand Reinhold, New York, 1971.
6. MORSE J. J. Dissolved air flotation in water treatment. *Wat. & Wat. Engng*, 1973, May, 161–163.

Chapter 8

1. IWAO M. Design and equipment outline of the Shiomidai purification plant. *J. Japan WatWks Ass.*, 1970, Oct., No. 433.
2. CULP G. L. and CULP R. L. *New concepts in water purification.* Van Nostrand Reinhold, New York, 1971.
3. PARKS SOUTHER G. and FORSELL B. O. The Lamella separator. Engineers Society of Western Pennsylvania, 1971.
4. WALTON R. and KEY T. D. Application of experimental methods to the design of clarifiers for waterworks. *J. Instn Civ. Engrs*, 1939, **13**, Nov., No. 1, 21–48.
5. Discussion. Application of experimental methods to the design of clarifiers for waterworks. *J. Instn Civ. Engrs*, 1940, **14**, Oct., No. 8, 485–495.

Chapter 9

1. HANSEN S. P. *et al.* Some recent advances in water treatment technology. *Chem. Engng Prog. Symp. Ser.*, 1969, **65**, No. 97, 207–218.
2. SMETHURST G. Some recent works in Bagdad. *J. Instn Wat. Engrs*, 1956, **10**, Mar., No. 2, 123.

Chapter 10

1. DEPARTMENT OF THE ENVIRONMENT. *EC directive relating to the quality of water intended for human consumption (80/778/EEC)*. DoE, London, circular 20/82.
2. DEPARTMENT OF THE ENVIRONMENT. Circular (letter) W17/1984. DoE, London.
3. WORLD HEALTH ORGANIZATION. *Guidelines for drinking-water quality—Vol. 1: Recommendation*. WHO, Geneva, 1984.

Chapter 12

1. COX C. R. *Operation and control of water treatment processes*. World Health Organization, Geneva, 1964, Chapter 12.
2. MÖRGELI B. *The removal of pesticides from drinking water*. Sulzer Brothers Ltd, Winterthur, Switzerland.

Chapter 13

1. COX C. R. *Operation and control of water treatment processes*. World Health Organization, Geneva, 1964.

Appendix 1

1. CROWE P. J. The Farmoor Source works in operation. *J. Instn Wat Engrs*, 1974, **28**, Mar., No. 2, 123–124.
2. CARTWRIGHT F. Design of Farmoor treatment works. *J. Instn Wat. Engrs*, 1964, **18**, Aug., No. 5, 401–411.
3. PUGH N. J. The treatment of doubtful waters for public supplies. *J. Instn Wat. Engrs*, 1957, **11**, Feb., No. 1, 24–31.
4. LAMONT J. Extensions to Hampton Loade works: Part 1—Engineering aspects. *J. Instn Wat. Engrs*, 1974, **28**, May, No. 3, 157–162.
5. HOLLAND G. J. Extensions to Hampton Loade works: Part 2—Chemical aspects. *J. Instn Wat. Engrs*, 1974, **28**, May, No. 3, 163–175.
6. ADAMS R. W. *et al.* The River Derwent scheme of the Nottingham Corporation. *J. Instn Wat. Engrs*, 1973, **27**, Feb., No. 1, 33–34.
7. GRIFFITHS J. H. T. and WISDISH W. J. L. Diddington treatment works, Great Ouse water supply scheme. *J. Instn Wat. Engrs*, 1967, **21**, Mar., No. 2, 106.

8. SAXTON K. J. H. Operation of the Grafham Water scheme. *J. Instn Wat. Engrs*, 1970, **24,** Oct., No. 7, 413.

9. MILNE J. W. The Pitsford treatment works. *J. Instn Wat. Engrs*, 1958, **12,** Mar., No. 2, 87–95.

10. ARDERN F. L. The Tittesworth Reservoir scheme of the Staffordshire Potteries Water Board. *J. Instn Wat. Engrs*, 1964, **18,** Mar., No. 2, 90–95.

11. PEPPER R. A. The use of activated silica as a coagulant aid at Wearhead pressure filtration plant. *J. Instn Wat. Engrs*, 1961, **15,** Feb., No. 1, 47.

12. BUCK A. C. The Johore River project, Singapore. *J. Instn Wat. Engrs*, 1969, **23,** Feb, No. 1, 16–19 and 37.

Glossary

1. HARDY F. S. and SMETHURST G. The River Adhaim flood of 1952. *J. Instn Wat. Engrs*, 1954, **8,** 145.

Conversion factors

Decimal multiples of units

Factor	Prefix	Symbol
10^3	kilo	k
10^{-1}	deci	d
10^{-2}	centi	c
10^{-3}	milli	m
10^{-6}	micro	μ

Length	inch foot	1 in = 25.40 mm 1 ft = 0.3048 m
Area	hectare square foot	1 ha = 10^4 m² 1 ft² = 0.092 90 m²
Volume	litre cubic foot UK gallon US gallon	1 l = 10^{-3} m³ 1 ft³ = 0.028 32 m³ 1 UKgal = 4.546 × 10^{-3} m³ 1 USgal = 3.785 × 10^{-3} m³

Velocity	foot per second
	UK gallon per square foot
	   per hour
	UK gallon per square
	   foot per day
		1 ft/s = 0.3048 m/s
		
		1 UKgal/ft² per h = 0.049 m/h
		= 1.175 m/d
		1 UKgal/ft² per day = 0.0489 m/d

Acceleration	foot per second squared standard acceleration due to gravity	1 ft/s² = 0.3048 m/s² g = 9.807 m/s²
Mass	pound grain tonne UK ton short ton (US ton)	1 lb = 453.6 g 1 gr = 0.064 80 g 1 t = 10^6 g 1 ton = 1.016 t 1 sh ton = 0.9072 t
Density	gram per cubic centimetre	1 g/cm³ = 1000 kg/m³
Concentration	part per million grain per UK gallon gram per cubic metre	1 ppm = 1 mg/l 1 gr/UKgal = 14.25 mg/l 1 g/m³ = 1 mg/l
Volume rate of flow	litre per second cubic foot per second UK gallon per hour million UK gallons per day	1 l/s = 0.001 m³/s 1 ft³/s = 0.028 32 m³/s 1 UKgal/h = 1.263 × 10^{-6} m³/s 1 mgd = 4546 m³/d
Force	newton gram-force	1 N = 1 kg m/s² 1 gf = 9.807 × 10^{-3} N
Pressure	pascal gram-force per square centimetre pound-force per square inch bar standard atmosphere	1 Pa = 1 N/m² 1 gf/cm² = 98.07 Pa 1 psi = 6895 Pa 1 bar = 10^5 Pa 1 atm = 1.013 × 10^5 Pa
Dynamic viscosity	centipoise gram per centimetre second	1 cP = 10^{-3} Pa s 1 g/(cm s) = 10^{-1} Pa s

Kinematic viscosity	centistoke stoke centimetre squared per second	$1 \text{ cSt} = 10^{-6} \text{ m}^2/\text{s}$ $1 \text{ St} = 10^{-4} \text{ m}^2/\text{s}$ $1 \text{ cm}^2/\text{s} = 10^{-4} \text{ m}^2/\text{s}$
Energy	joule kilowatt hour horsepower hour British thermal unit	$1 \text{ J} = 1 \text{ N m} = 1 \text{ W s}$ $1 \text{ kW h} = 3.6 \times 10^6 \text{ J}$ $1 \text{ hp h} = 2.685 \times 10^6 \text{ J}$ $1 \text{ Btu} = 1055 \text{ J}$
Power	watt foot pound-force per second horsepower	$1 \text{ W} = 1 \text{ J/s}$ $1 \text{ ft lbf/s} = 1.356 \text{ W}$ $1 \text{ hp} = 745.7 \text{ W}$

Temperature

°C	°F	Notes
0	32	Water freezes
10	50	—
20	68	—
30	86	—
100	212	Water boils

Index